KU-756-736

D. A. FRASER

The Physics of
Semiconductor Devices

FOURTH EDITION

CLARENDON PRESS · OXFORD

1986

Oxford University Press, Walton Street, Oxford OX2 6DP

Oxford New York Toronto
Delhi Bombay Calcutta Madras Karachi
Petaling Jaya Singapore Hong Kong Tokyo
Nairobi Dar es Salaam Cape Town
Melbourne Wellington
and associated companies in
Beirut Berlin Ibadan Nicosia

Published in the United States by
Oxford University Press, New York

© D. A. Fraser, 1977, 1979, 1983, 1986

All rights reserved. No part of this publication may be reproduced,
stored in a retrieval system, or transmitted, in any form or by any means,
electronic, mechanical, photocopying, recording, or otherwise, without
the prior permission of Oxford University Press

This book is sold subject to the condition that it shall not, by way
of trade or otherwise, be lent, re-sold, hired out or otherwise circulated
without the publisher's prior consent in any form of binding or cover
other than that in which it is published and without a similar condition
including this condition being imposed on the subsequent purchaser

First edition 1977
Second edition 1979
Third edition 1983
Fourth edition 1986

British Library Cataloguing in Publication Data
Fraser, D. A.
The physics of semiconductor devices.—
4th ed.—(Oxford physics series)
1. Semiconductors
I. Title
537.6'22   QC611
ISBN 0–19–851867–6
ISBN 0–19–851866–8 Pbk

UNIVERSITY OF BRADFORD
LIBRARY
20 NOV 1986

ACCESSION No.   CLASS No.
0288385 01M 537·311·33 FRA

LOCATION

Library of Congress Cataloging in Publication Data

Fraser, D. A.
The physics of semiconductor devices.
(Oxford physics series; 16)
Bibliography: p.
Includes index.
1. Semiconductors. I. Title. II. Series.
QC611.F72   1986   537.6'22   86-5325
ISBN 0–19–851867–6 √
ISBN 0–19–851866–8 (pbk.)

Typeset and printed in Great Britain by
The Universities Press (Belfast) Ltd.

Editor's Foreword

Physics would be paralysed without electronic devices. An understanding of what these devices can and cannot do is essential for the systems engineer and the computer scientist. The physical principles on which such devices depend need to be known by physicists, electronicists, electrical engineers, and many chemists and material scientists.

The treatment is basically that of a second-year degree course, but some teachers may prefer to present part of the material in the third year rather than confine it to an integrated course. This volume is complete in itself as far as the subject matter allows.

The first two chapters of the book define and develop the relevant concepts and relations, before considering individual devices. Wherever possible, the theory is presented in a sufficiently general way to make it of lasting value, not limited by the patented semiconductor products at present in use. The methods are developed with the aid of many diagrams, which often approach the problems from alternative points of view. Manufacturing processes are not treated in undue detail, but the industrial significance of demands for reliability, low noise, and other desirable features is recognized and evaluated.

Overlap with other texts in the Oxford Physics Series is limited and serves to show the need for any applied topics to be based on an adequately wide and thorough understanding of physics. *Atoms in contact* (OPS 5, Jennings and Morris) and *D.C. and A.C. circuits* (OPS 2, Lancaster) will give help in the study of semiconductor physics and Rosenberg's *The solid state* (OPS 9) has obvious relevance.

E.J.B.

Preface

This book presents the basic physical processes of semiconductor devices, and relates these to the functions that the devices perform. The material that is included has been presented to students in courses ranging from first-year undergraduate to M.Sc., but the general level is that of a second-year undergraduate. The treatment is mathematical only wherever useful results can be quickly derived, and relies on diagrams to a considerable extent. The book has been kept as compact as possible, so that many topics are treated to a useful depth in a fairly short space.

I have been encouraged over the last few years by reports of earlier editions being seen on working scientists' bookshelves, so that it is apparent that the readership extends beyond the electronics and physics students for whom it was written. At the same time, students can take assurance that the topics they study in these chapters are found to be of use in practical situations.

New material in this edition covers some of the recent advances in the very active topic of physical electronics that are either giving promise of becoming important, or that have already led to major applications. Preparation and use of 3–5 compound semiconductors and of multiple thin layers are discussed, and devices for optical fibre communications are now described. In addition there are minor revisions at many points.

Once again it is a great pleasure to acknowledge the assistance and encouragement provided by the author's colleagues, students, and family in the many aspects of the study of this subject.

Kings College (KQC), London, 1985   D.A.F.

Contents

1. Electrons in solids

Reasonable questions to ask about solids that conduct electricity are: where are the electrons that carry the current? How many are there? How quickly and in what directions are they going? Some of these questions have straightforward answers, which are given in this chapter; others take much time and paper, and we can only move part of the way to a full answer.

Four descriptions of solids are commonly used in discussing semiconductors. The most sweeping—the equivalent-circuit description—deals only with relations between the terminal voltages and currents in a complete device. Such a description can be regarded as a summary of the understanding gained by the more physical approaches. Of these, the bond model is often too simple to be useful, the energy-band model will be frequently used, and energy/wave-number diagrams will occasionally be needed to explain more subtle processes.

The bond model

Most semiconductor devices are made from crystals, and we shall concentrate on these.

Although crystals occur with a marvellous variety both in their external form and the internal patterns of arrangement of their atoms, nearly all important semiconductors have the same crystal structure, that of diamond. From a crystallographer's point of view this is a face-centred cubic structure with eight atoms in the unit cell. A consequence is that a semiconductor is isotropic for many important physical properties, i.e. the property has the same value in all directions. The atoms are arranged so that round any and every atom there are four others making a

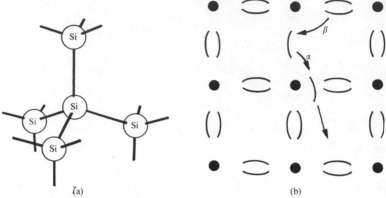

Fig. 1.1.   Bonds: (a) each atom in a Si crystal is bound to four others; (b) each bond includes two electrons. At α, an electron has been freed from the bond, allowing it to move, and allowing other transitions, such as β, to occur.

tetrahedron in space (shown in the case of a Si crystal in Fig. 1.1(a)). In a study of semiconductors this tetrahedral arrangement is very important.

There are three important kinds of direction in a cubic crystal. The directions of the cube edges are called (100)-directions by crystallographers, the directions from the centre of the cube to the middle of each of the 12 edges are (110)-directions, and from the centre to the eight cube vertices are (111)-directions. It is of interest to look at a model of a diamond-structure crystal to see in which planes the atoms are most densely packed.

The average positions of the electrons relative to the atomic cores can be pinned down by calculation and by X-ray measurements. What one finds is that all except four of the electrons stay close to the nucleus of the atom; the last four, however, tend to be found in the regions directly between two atoms. These four together with the equivalent electrons from neighbouring atoms form the covalent bonds which hold the crystal together, and provide from their number the mobile charges which carry current. The multiple tetrahedral arrangement is not easy to draw, and a flattened pattern is simpler.

In Fig. 1.1(b) one electron is shown displaced from its proper position. Such an electron can move through the crystal and thus carry current. In addition there is a space where it used to be. That space cannot in itself carry current, but it allows other electrons the opportunity of moving without having to be given enough energy to break out of their bond. As another electron moves into the space, it leaves another space behind it, and so on. It is easier to focus attention on the moving space for an electron rather than a succession of different electrons. The moving space is usually called a hole.

Thus in the bond picture electrical currents can be carried by both electrons and holes simultaneously.

Energy bands

First we shall explain what is meant by energy bands, then discuss why nature works this way.

Each electron in a solid has a certain total amount of energy, made up of kinetic and potential energy. In a particular sample of a solid there are some ranges of total electron energy where electrons can be found, and other ranges where there are no electrons (Fig. 1.2(a)). Close investigation shows that at fairly high values of total energy, there are ranges where, though electrons are rare, they can occur. The complete energy axis can thus be divided into

   (a) forbidden bands—no electrons have these energies; and

   (b) allowed bands—there may be electrons at these energies.

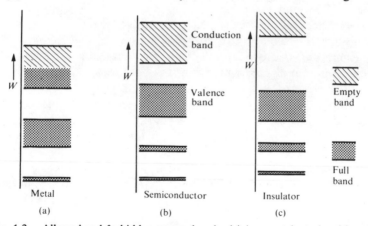

Fig. 1.2.   Allowed and forbidden energy bands; (a) in a metal one band is only partly filled; (b) in a semiconductor, the valence band is full (nearly) and the conduction band is empty (nearly); (c) the energy gap between valence and conduction band is larger in an insulator than in a semiconductor.

Usually the lower bands are full of electrons and the upper ones empty. The allowed and forbidden bands come about because

(a) electrons have a wave-like nature:

(b) if the wave equations describing the electrons are to have suitable solutions, some parameters (quantum numbers) must have special values;

(c) (which follows from (b)) an electron has to occupy a 'state', as characterized by a set of quantum numbers;

(d) the regular arrangement of atoms in a crystal leads to electron states having energies only within certain ranges.

For many problems, the band picture used without energy values is very helpful. For instance, Fig. 1.2 shows how metals, where the highest-energy electrons are in the middle of a band, are different from semiconductors and insulators, in which one band, called the *valence band,* is full, and the next band up, the *conduction band,* is empty. The magnitude of a forbidden energy band is known as the *band gap* between the two allowed bands above and below it. The band gap usually referred to is that between the valence and conduction bands: its width is usually measured in electronvolts (eV).

The distinction between insulators and semiconductors is one of degree rather than kind—insulators have larger band gaps, perhaps 3 eV or more, while semiconductors have band gaps from 2·5 eV down to 0·1 eV.

One general point should be noted about energy-band diagrams. The energy axis has plus at the top and minus at the bottom, but because an electron has a negative charge, the electric potential axis, which can also be drawn, has minus at the top and plus at the bottom of the page: be wary.

From a quantum-mechanical point of view each band is finely subdivided, a semiconductor crystal of N atoms having $4N$ electron states in the conduction band and $4N$ states in the valence band. N may be as large as 10^{14} for a small transistor, so the subdivisions are very fine. For a more detailed treatment of this point look up the Kronig–Penney model or Mathieu's equation in books on solid-state physics (e.g. Rosenberg, OPS 9).

Having considered the possible states that an electron is allowed to occupy, the next question is which states are occupied? Pauli's exclusion principle says that no two electrons may occupy the same state. At absolute zero temperature the electrons would go into the lowest energy states available. At room temperature, there is a little more energy available for distribution among the electrons, but it is only enough for a small proportion of the electrons in the uppermost levels to be excited a little. For an exact analysis we need to know how many electrons there are at each energy; this is discussed on p. 11.

Energy/wave-number diagrams

This level of description takes account of the momentum, and hence the direction of travel, of the electrons as well as their total energy. It is based on a wave picture for the properties of the electrons, and this is derived from wave mechanics (also known as quantum mechanics or matrix mechanics). The wave provides a way of calculating the effects an electron can produce; we can say if we like that the wave is the electron.

Avoiding mathematics, some useful results are:

$$\text{energy} \qquad W = \hbar\omega; \qquad \frac{h \cdot 2\pi f}{2\pi} = hf \qquad (1.1)$$

$$\text{momentum} \qquad p = \hbar k; \qquad (1.2)$$

$$\text{group velocity} \qquad v = \frac{d\omega}{dk} = \frac{dW}{dp}, \qquad (1.3)$$

where $\hbar$ = Planck's constant $(h)/2\pi$, and ω is the radian frequency of the wave. Equation (1.1) is familiar from such examples as the photoelectric effect. Equation (1.2), relating the momentum of a particle (or a wave) to its wave number k (the number of radians of phase change in unit distance), is less often seen, though it is of equal status to eqn (1.1). Equation (1.3) is the one to use when analysing the motion of a wave packet, which is the way we describe a particle by using waves.

In a solid, the boundaries reflect waves, and if destructive SWR interference is not to occur, the wavelengths must have special values. We look at a simple one-dimensional model, a row of N atoms distance a apart, each with two free electrons, and see what values of k can be used to make allowed wavefunctions, and hence states.

The longest wavelength that can fit our one-dimensional

$$\frac{\lambda}{2} = Na \implies \lambda = 2Na$$

$$\lambda \neq$$

$$\frac{2\pi}{\lambda} = k$$

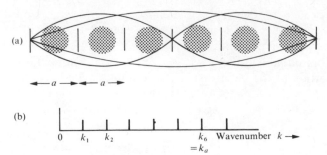

Fig. 1.3. Waves in a line of six atoms: (a) the longest two wavelengths are $12a$ and $6a$; (b) the wave numbers of the allowed waves are evenly spaced along the k axis.

'crystal' is $2Na$, so the lowest k is given by $\pi/Na = k_1$. The next wave that can fit the boundary conditions has one complete wavelength in the crystal, so that $k_2 = 2\pi/Na = 2k_1$. After that, $k_3 = 3 \times \pi/Na$, and so on, as in Fig. 1.3, until $k_{2N} = 2N \times \pi/Na = 2\pi/a = k_a$. We see that all the allowed values of k are equally spaced, and that $2N$ of them fill the range from 0 to k_a independent of the size of the crystal, and hence the value of N.

The regular spacing of allowed values of k occurs in three directions when a three-dimensional lattice of atoms in ordinary space is analysed. Thus to obtain the view of electron states most appropriate for the start of mathematical analysis, one does well to work in k-space or momentum space. The $2N$ waves, with k running from k_1 to k_a, are all the waves one needs to make a band. Notice that they are just enough for the number of electrons in the crystal, and that the wavelength goes from the size of the crystal to the distance between atoms.

If one tries to use $k > k_a$, one does not get a new independent solution to the wave equation. This is stated by Floquet's theorem—the Bloch principle. One is allowed to use $k > k_a$ if it is convenient, but the state with wave-number $k(>k_a)$ is the same as the state which has a wave-number k_a lower.

For another band, other quantum properties of the state have changed, so we may use the same wave numbers again. The energy of an electron as a function of its wave-number is sketched in Fig. 1.4(a) for a one-dimensional system. To be precise about electron energy is much more difficult than to be

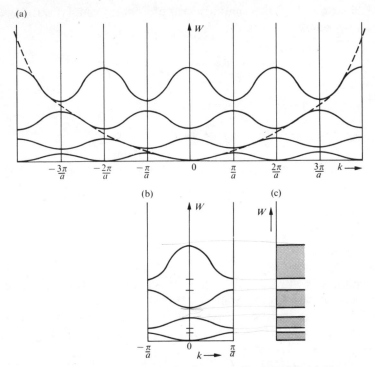

Fig. 1.4.   Electron energy/wave-number in a periodic potential. (a) W repeats at regular intervals of k; (b) all the information is in the range $\pm\pi/a$; (c) the related energy band diagram.

precise about momentum, so Fig. 1.4(a) shows only the sort of shape that the curves might have. Remember that the curves should be dotted lines with the dots spaced equally along the k axis rather than continuous, each dot giving the energy and momentum for a state. One can see that the energy is unaltered if we add k_a to any k—the figure is periodic along the k axis, because the real crystal is periodic in space.

If just one interval of k, of width k_a (a Brillouin zone) is shown as in Fig. 1.4(b), the figure is known as the reduced zone diagram. It still contains all the information of Fig. 1.4(a), the extended zone diagram.

The allowed and forbidden energy ranges are clear in Fig. 1.4(c). In a real semiconductor, we have to specify which direction in the crystal k is along. The $W - k$ curves are different for different directions, and are also of a more complex structure than our simple figure.

Effective mass

Referring back to eqn (1.3), we can write for the acceleration f of an electron due to some external force

$$f = \frac{dv}{dt} = \frac{d}{dt}\left(\frac{dW}{dp}\right) = \frac{dp}{dt} \cdot \frac{d}{dp}\left(\frac{dW}{dp}\right).$$

Thus

$$f = \frac{dp}{dt} \cdot \frac{d^2W}{dp^2}.$$

But dp/dt is the rate of change of momentum, and hence equals the applied force P. Thus $(d^2W/dp^2)^{-1}$ replaces the mass in the equation of motion $P = mf$, and we can describe the response of a carrier to a force by using $(d^2W/dp^2)^{-1}$ instead of the mass. This new term is known as the *effective mass* (m^*) of a carrier, and summarizes the way interaction with the lattice affects the carrier motion. Explicitly,

$$m^* = \left(\frac{d^2W}{dp^2}\right)^{-1} = \hbar^2\left(\frac{d^2W}{dk^2}\right)^{-1}. \tag{1.4}$$

The effective mass of an electron in the bottom of the conduction band is usually less than the free-electron mass, and may be much less.

The curvature of the top of the valence band can be seen to be of the opposite sign to that of the bottom of the conduction band. The effective mass of an electron at the top of the valence band is therefore negative, though it is more usual to consider holes near the top of the valence band, with effective mass and charge both positive.

The $W - k$ diagrams for Ge, Si, and GaAs that are equivalent to Fig. 1.4 for the ideal semiconductor are shown in Fig. 1.5. Notice that the two halves of each diagram do not match; the left part describes directions along (111)-directions while the right part refers to (100)-directions. There are several bands shown for

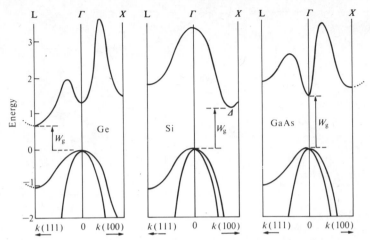

Fig. 1.5.   The band structure of Ge, Si, and GaAs in two important directions. The bands are horizontal where they meet the zone boundaries. In the valence bands of each substance there are both light and heavy holes.

each substance, but the important features all border on the forbidden-energy gap.

For all three materials there are two valence bands, which have different values of d^2W/dk^2 and, therefore, different effective masses. A sharply curved band has a large value of d^2W/dk^2 and hence a small effective mass.

The curvature of a light hole band along (111) is less than along (100). As a result the effective mass varies with direction. Although this difference is normally not important, it forms the basis of p-type silicon strain gauges, discussed in Chapter 2.

The first point to note about the conduction band in each part of Fig. 1.5 is the height of its lowest point above the highest point of the valence band. This is a measure of the forbidden band or 'band gap', and is the single number which is taken as the most effective assessment of a semiconductor. Large band gaps are essential for high-temperature operation, for high resistance, and for the emission of visible light. Low band gaps offer a chance to detect low-energy infra-red photons or to make low-voltage devices.

The relative position along the axes of the valence-band maximum and the conduction-band minimum is also important. When the two are at the same k-value, as in GaAs at $k = 0$, then

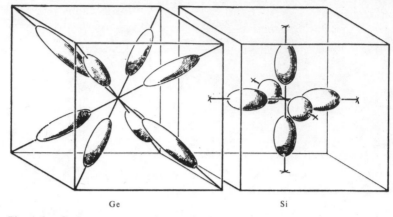

Ge   Si

Fig. 1.6.   Constant-energy surfaces near the conduction-band minima in Ge and Si. (Taken by permission from Ziman (1960). *Electrons and phonons*. Oxford University Press.)

the semiconductor is classified as having a *direct gap*. Direct-gap semiconductors are used to make light-emitting diodes, which are described in Chapter 4. The second minimum in the GaAs conduction band is usually empty, but can be populated in the conditions of use of a Gunn diode (Chapter 2). Germanium and silicon have the minimum of the conduction band at a different value of k from the maximum of the valence band. They are *indirect-gap* semi-conductors. The letters L, Γ, X, and Δ are used in theoretical solid state physics to indicate special points in a Brillouin zone.

Figure 1.6 shows constant energy surfaces in the regions of k-space near the conduction-band minimum for Ge, Si, and GaAs. These are the regions where the states will be occupied. In Ge, each L minimum in a (111)-direction is on the Brillouin zone boundary, so there are in effect four equivalent minima in the Ge conduction band. In Si, the six Δ minima are all in the first Brillouin zone. In Si and Ge, the constant energy surfaces are ellipsoids of revolution. The radius of curvature in different directions, and hence the effective mass is different in different directions. A full mathematical treatment uses a tensor effective mass. Measurable properties of a device usually depend on some average of the effective masses of the electrons involved. The appendix quotes values that are correct for the movement of carriers in an electric field. For GaAs, the constant energy surface is well described by a sphere centred at $k = 0$.

Density of states

We can put more detail into our picture of semiconductors by describing the number of electrons at each energy. This will take two steps, first describing the number of states per unit energy range per unit volume—the density of states, and then the fraction at each energy that are filled.

The density of states $S(W)$ as a function of total electron energy W is shown in Fig. 1.7 for a semiconductor. $S(W)$ is zero in the band gap, and has a calculable shape elsewhere. Free electrons and holes occupy states near the band edges, and the shape of $S(W)$ near the band edges is simple, consisting of parabolas springing at right angles from the energy axis. The states fill k-space, being uniformly spaced in three dimensions in the same way as is shown in Fig. 1.3 for one dimension. The states with a value of k between k and $k + dk$ fill a spherical shell

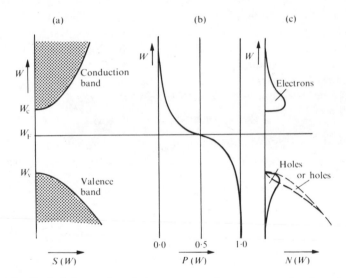

Fig. 1.7.   To find the number of electrons per unit energy per unit volume in (c), multiply the number of states per unit energy per unit volume in (a) by the probability of the states being occupied in (b). The number of holes is the number of states not occupied, and equals $\{1 - P(W)\}S(W)$.

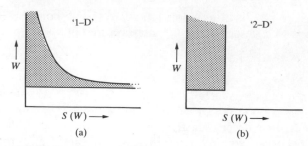

Fig. 1.8. Distribution of states for electrons. (a) In a one-dimensional system. (b) In a two-dimensional system. For comparison with Fig. 1.7 and Fig. 5.21.

of volume

$$S'(k)\,\mathrm{d}k = \frac{\mathrm{d}}{\mathrm{d}k}\,(\tfrac{4}{3}\pi k^3)\,\mathrm{d}k = 4\pi k^2\,\mathrm{d}k.$$

Since we have $p = \hbar k$ and $W = W_\mathrm{c} + (p^2/2m_\mathrm{e})$ for electrons, then we can obtain

$$S(W)\,\mathrm{d}W = \frac{4\pi}{\hbar^3}\,(2m_\mathrm{e})^{\frac{3}{2}}(W - W_\mathrm{c})^{\frac{1}{2}}\,\mathrm{d}W \tag{1.5}$$

For two- and one-dimensional lattices, the allowed states fill a circular strip of width $\mathrm{d}k$ and radius k, and points on a line between k and $k + \mathrm{d}k$ as in Fig. 1.8. The two- and one-dimensional distributions are relevant in the thin layers used in modern devices.

For holes in the upper edge of the valence band,

$$S(W) = (4\pi/\hbar^3)(2m_\mathrm{h})^{\frac{3}{2}}(W_\mathrm{v} - W)^{\frac{1}{2}}. \tag{1.6}$$

W_c is the energy of the lower edge of the conduction band, W_v that of the upper edge of the valence band, m_e and m_h are the effective masses of carriers in the two band edges, and h is Planck's constant. Notice that m_e and m_h are the only terms in eqns (1.5) and (1.6) which refer to a given substance.

Fermi–Dirac statistics

Electrons are found to have the fundamental properties that they are indistinguishable from one another, and that each must be in a distinguishable state. This latter point is known as the Pauli exclusion principle. Particles which have these properties are known as fermions, and are described by Fermi–Dirac statistics. Hence for a system in thermal equilibrium,

$$P(W) = 1/[1 + \exp\{(W - W_F)/\kappa T\}]. \tag{1.7}$$

$P(W)$ is the probability that a state at an energy W is occupied by an electron, κ is the Boltzmann constant, and T is the temperature of the system. We can find where the energy W_F, the Fermi energy or *Fermi level* must be by noticing that when $W = W_F$, $P(W)$ must equal $\frac{1}{2}$. A suggested definition for W_F is *'The Fermi level is that energy where the chance of a state being occupied by an electron is one half'*.

The Fermi levels can come in a band gap, but we can do our sums on the chance of a state being filled, even though there are no states at that particular energy. Fig. 1.7(b) shows a graph of $P(W)$. Notice the way it is symmetrical about the $\{W = W_F, P(W) = \frac{1}{2}\}$ point, and tends to zero for W very positive, and to one for W very negative. If W is usefully greater than W_F, then there is a simpler approximate form of $P(W)$, because $\exp\{(W - W_F)/\kappa T\} \gg 1$.

$$P(W) \sim 1/[\exp\{(W - W_F)/\kappa T\}] = \exp\{-(W - W_F)/\kappa T\}. \tag{1.8}$$

This form can be derived by assuming that electrons behave like classical micro-billiard balls, and leads to the Maxwell–Boltzmann distribution, but its justification is that for $W \gg W_F$ it is a good approximation to eqn (1.7), derived from non-classical Fermi–Dirac statistics.

To know how many holes there will be in a sample we must know what fraction of states are unoccupied, i.e. we must know $1 - P(W)$

$$1 - P(W) = 1 - 1/[1 + \exp\{(W - W_F)/\kappa T\}]$$

$$1 - P(W) = \exp\{(W - W_F)/\kappa T\}/[1 + \exp\{(W - W_F)/\kappa T\}].$$

When we are interested in holes, the relevant energy range is often well below W_F, and $\exp\{(W - W_F)/\kappa T\} \ll 1$. Then

$$1 - P(W) \sim \exp\{(W - W_F)/\kappa T\}. \tag{1.9}$$

Notice that eqn (1.9) is like eqn (1.8) with a sign reversed. This fits with the idea that holes carry a positive charge, as their energy should then increase as the potential becomes more positive.

Electron and hole densities

We now have a formula for both $S(W)$ and $P(W)$, and can find the number of electrons $N(W)$ actually occupying states at various energies (Fig. 1.7(c)):

$$N(W) = S(W) \times P(W).$$

We can describe the number of electrons near the bottom of the conduction band by using the expression for $S(W)$ from eqn (1.5) and the approximate formula for $P(W)$ (eqn (1.8)):

$$N(W) = \frac{4\pi}{h^3}(2m_e)^{\frac{3}{2}}(W - W_c)^{\frac{1}{2}}\exp\{-(W - W_F)/\kappa T\}.$$

The total number of electrons per unit volume over some definite energy range is found by integrating $N(W)$. The number n of electrons in the conduction band is given by

$$n = \int_{W_c}^{W_t} N(W)\,dW,$$

where we integrate from W_c, the energy of the lower edge of the band to W_t at the top of the band. This integral turns out to be awkward, but if W_t is replaced by $+\infty$, the answer is essentially the same, and the answer comes in a standard form:

$$n = \int_{W_c}^{\infty} N(W)\,dW$$

$$= \frac{2}{h^3}(2\pi m_e\kappa T)^{\frac{3}{2}}\exp\left\{-\frac{(W_c - W_F)}{\kappa T}\right\} \text{ electrons per unit volume.}$$

$$(1.10)$$

The number of holes p in the valence band is found in the same way, using $1 - P(W)$ instead of $P(W)$ and integrating down from W_v at the top of the valence band:

$$p = \frac{2}{h^3}(2\pi m_h\kappa T)^{\frac{3}{2}}\exp(-(W_F - W_v)/\kappa T) \text{ holes per unit volume.}$$

$$(1.11)$$

If the terms outside the exponentials in (1.10) and (1.11) are collected into single symbols A_c and A_v, where

$$A_c = 2(2\pi m_e \kappa T)^{\frac{3}{2}}/h^3,$$
$$A_v = 2(2\pi m_h \kappa T)^{\frac{3}{2}}/h^3,$$

(1.12)

then the expressions for n and p look more compact.

$$n = A_c \exp\{-(W_c - W_F)/\kappa T\}$$
$$p = A_v \exp\{-(W_F - W_v)/\kappa T\}.$$

(1.13)

A_c and A_v are called the effective density of states for the conduction band and for the valence band. They can be thought of as the number of states that would be required to give the same value of n (or p) if all the states were at a single energy, that of the band edge. From eqn (1.12) we can see that A_c and A_v vary with temperature, while $S(W)$, the density of states, varies very little.

W_F appears in both the eqns (1.13). It must be the same value both times, so we can solve for W_F or eliminate it.

Multiplying the two parts of (1.13) we get

$$pn = A_c A_v \exp\{-(W_c - W_v)/\kappa T\}.$$

The right-hand side of this equation is dependent only on the temperature and the kind of semiconductor, and not on hole or electron densities, and therefore the left-hand side must be the same value for differently doped (see p. 16) samples of the same semiconductor. In intrinsic semiconductors without added impurities, the electron density equals the hole density, and the name *intrinsic carrier density*, n_i, is given to this carrier density. Thus

$$pn = n_i^2.$$

(1.14)

This equation is very important in the study of semiconductors, but seems to be difficult to emphasize adequately. For this reason it is suggested that it be called the *semiconductor equation*. It is an example of the chemical law of mass action. The second result from eqn (1.13) comes from dividing the two equations:

$$p/n = (A_v/A_c) \exp\{(W_c + W_v - 2W_F)/\kappa T\}$$

or $\qquad\qquad\qquad\qquad\qquad\qquad\qquad\qquad\qquad$ (1.15)

$$W_F = (W_c + W_v)/2 - \frac{\kappa T}{2} \ln(p/n) - \tfrac{3}{4}\kappa T \ln(m_e/m_h).$$

Equation (1.15) shows that the Fermi level for intrinsic material (where $\ln(p/n) = 0$) is at the average of W_c and W_v, with a small correction when $m_e \neq m_h$. When $p \neq n$, then the Fermi level shifts towards the band with the majority of carriers.

It is worth remembering that the semiconductor equation holds for any values of p or n, as long as a thermal equilibrium situation is being described, so it often is valid. However, when p and n are controlled in some device by the external conditions, then we may be far from thermal equilibrium and if so

$$n_i^2 \neq p_{\text{noneq}} n_{\text{noneq}}$$

Doping and carrier density

Many semiconducting devices are made by introducing small quantities of the proper kind of chemical impurity into the semiconductor lattice—doping the semiconductor.

There are two kinds of dopant, donors and acceptors. A number of pairs of terms that go with donors or acceptors are listed.

Type of dopant	Donor (N_d)	Acceptor (N_a)
Semiconductor type	n-type	p-type
Majority carriers are	electrons	holes
They occur in the	conduction band	valence band
Minority carriers are	holes	electrons
Dopants on Si or Ge	P, As, Sb	B, Al, Ga
	group 5	group 3
Dopant in GaAs etc.	Se, Te	Si
on group 5 site	group 6	group 4
Dopant in GaAs etc.	Ge, Si	Zn
on group 3 site	group 4	group 2
Charge on ionized dopant is	+ve	−ve
Fermi level is nearer	conduction band	valence band

The extra electron attached to a donor can be liberated by a small amount of energy, 10–50 meV, so that at room temperature, where $\kappa T \sim 25$ meV it is free to travel over the whole crystal and contribute to an increased conductivity. Acceptors remove an electron from the valence band, leaving a mobile hole.

If both donors and acceptors are added to the same crystal, their effects cancel—only the excess of one above the other has an effect on the free-carrier densities. This is exploited in device manufacture as an n-type semiconductor can be turned into p-type by adding an excess of acceptors. Indeed the process can be repeated, though three changes between n-type and p-type is about the limit.

If the doping atoms are all ionized, then in an electrically neutral region

$$n + N_a = p + N_d, \tag{1.16}$$

where N_a and N_d are the acceptor and donor number density.

Equation (1.16) in conjunction with $pn = n_i^2$ is sufficient to fix p and n if N_d and N_a are known. If $|N_a - N_d| \gg n_i$ then there are simple solutions. If $N_a > N_d$, the material is p-type, $p = (N_a - N_d)$, and $n = n_i^2/(N_a - N_d)$. If $N_d > N_a$ then the material is n-type, $n = (N_d - N_a)$ and $p = n_i^2/(N_d - N_a)$. It is seldom necessary to deal with situations when $|N_a - N_d| \gg n_i$, but if these occur, approximate values for p and n can be obtained by first ignoring the minority-carrier density in (1.16), and then using eqns (1.14) and (1.16) alternately to obtain more accurate solutions. Notice that in doped material, the minority-carrier density is much less than n_i.

3–5 Semiconductors

Si and Ge crystals have the diamond structure. This is modified in the 3–5 compound semiconductors only in that alternate atoms are from group 3 and from group 5. Thus a group 3 atom is surrounded by four group 5 atoms, and a group 5 atom is surrounded by four group 3 atoms.

The elements that are most used are in the second, third, and fourth periods of the periodic table of elements: group 3 Al, Ga,

In, and group 5 P, As, and Sb. Such compounds as BN show semiconducting properties, but are less used.

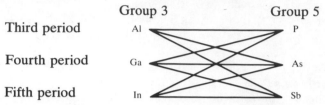

There are thus nine binary 3–5 semiconductors to consider. Their properties are listed in the Appendix. The central compound is GaAs, and this is the most important 3–5 semiconductor, being most studied, and having more uses than any other semiconductor apart from Si. The nine 3–5 binary compounds that have been referred to offer a useful range of semiconductor properties, but this range can be extended if ternary compounds are considered. These are compounds of three elements, where the ratio of group 3 atoms to group 5 atoms is held at 1:1, but one of the two classes of atom (3 or 5) is made up of a mixture. An example is $Al_{0.3}Ga_{0.7}As$, where the ratio of Al atoms to Ga atoms is 30:70 but there are equal numbers of group 3 atoms (Al, Ga) and group 5 atoms (As). The general formula for this kind of compound is written $Al_xGa_{1-x}As$. If there is a mixture of two group 5 elements, the formula might be GaP_yAs_{1-y}.

Figure 1.9 shows the relation between the band gap and the lattice constant for twelve sets of ternary alloys. The diagram can be viewed as a distortion of a simple pattern of four squares, with GaAs in the centre. There are only twelve alloys shown rather than $6 \times 3 = 18$, as the substitution of a third period element by a fifth period element (e.g. In for Al) is more difficult than the substitution of a fourth period element for one of its neighbours. A dashed line separates indirect gap materials at the top of the diagram from direct gap materials lower down. Over some parts of the diagram, the crystals will not form from the liquid mixtures: there is a miscibility gap.

By choosing the right proportions for the mixture, a semiconductor with any wanted band gap may be synthesized. There is still one problem left. The lattice constant of an epitaxial layer has to match closely that of the substrate if the epitaxial layer is to grow. Thus Fig. 1.9 shows that it is possible to grow $Al_xGa_{1-x}As$ on GaAs for all x, as the lattice constants vary only

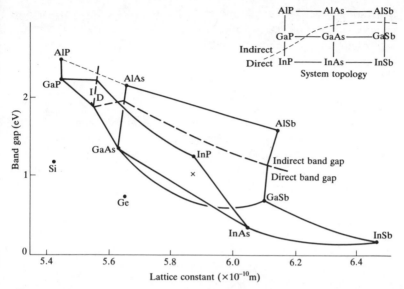

Fig. 1.9. The variation of band-gap with composotion for some ternary 3–5 semiconductors. The AlP–AlAs line is a likely guess. The X marks the properties of a laser material discussed in the text.

little, but it is difficult to find a substrate for $InAs_ySb_{1-y}$. The difficulty can be overcome by the use of quaternary alloys, using four elements, and still keeping the group 3:group 5 ratio = 1:1. The extra degree of freedom allows a designer to choose the band-gap that is wanted while keeping the lattice constant equal to that of a convenient substrate—probably GaAs or InP.

There are three ways of using four elements to make a 3–5 compound: 3,3–5,5; 3,3,3–5; and 3–5,5,5. A 3,3–5,5 quaternary compound can be thought of as lying in a region enclosed on Fig. 1.9 by four quaternary compound lines. Thus $In_xGa_{1-x}As_yP_{1-y}$ is in the region bounded by the lines joining GaAs, GaP, InP, and InAs.

Take an example: $In_{0.75}Ga_{0.25}As_{0.44}P_{0.56}$ is grown on an InP substrate as the active region of a semiconductor laser emitting photons of energy 1·0 eV. The lattice match to InP defines the lattice constant, and the photon energy defines the band gap, so the compound can be plotted on. Fig. 1.9. If a higher energy photon was wanted, then the ratio of P:As would be increased, and the ratio of Ga:In decreased to bring the lattice constant back to a match to InP.

In summary, 3–5 semiconductor band-gap engineering works by picking a substrate, and then finding a suitable quaternary compound that matches the lattice constant of the substrate and has the required band-gap.

Doping Methods

Doping atoms can be introduced into the semiconductor crystal in several ways. In the growing and purification of the original crystal it is convenient to arrange that its properties will be useful without further doping, at least for part of a device.

One way of doping the whole of a block of silicon uses the transmutation of elements. When a sample of undoped Si is placed in the neutron flux from a nuclear reactor, transmutation of some of the Si nuclei into P nuclei occurs. The phosphorus is in group 5 of the periodic table, so the result is to produce n-type silicon. The nuclear reaction is described as

$$\mathrm{Si}_{14}^{30}(n, \gamma)\mathrm{Si}_{14}^{31}(-, \beta^-)\mathrm{P}_{15}^{30}.$$

The new phosphorus atoms are distributed more evenly than those introduced in other ways, so that manufacturers of integrated circuits need to allow less tolerance in their designs for irregularities in the doping density of the starting material (the substrate) when using transmutation doped silicon.

A very common method of introducing donor atoms is by diffusion at high temperature from the surface of a sample. We note here that diffusion plays an important part in the physics of semiconductor devices. The doping atoms diffuse to their required positions, charges diffuse while the device is operating, and heat is carried to the heat sink by thermal conduction—another example of diffusion.

Thermal diffusion has been much analysed and one can often take solutions from thermal problems and apply them to the other types of diffusion.

The ease with which a doping atom diffuses into a crystal may be described at different temperatures by a temperature-dependent coefficient of diffusion D:

$$D = D_0 \exp(-\Delta W / \kappa T).$$

The term ΔW is the energy barrier an atom has to surmount on each step of its diffusing journey. It is usually a few electronvolts. Diffusion is carried out at high temperatures—around 1000 K for Si. An hour or so is usually long enough for a stage in a manufacturing process.

In general, a brief diffusion at high temperature will produce a shallow layer, but a long diffusion at a lower temperature is required to diffuse doping atoms deep into a piece of semiconductor. The doping atoms may be supplied to the surface of the sample by a gaseous compound containing the wanted atoms—AsH_3, BH_3, and $AlCl_3$ are examples—or by depositing a solid layer of some substance containing the doping atoms on the surface, and then removing the unwanted excess after the diffusion process has been completed.

Another method of introducing doping atoms is by ion implantation. A beam of ions of energy between 50 keV and 500 keV is fired at the surface. The depth of penetration increases up to 0.5×10^{-6} m as the beam voltage increases, so that the distribution and doping density are readily controllable, though elaborate equipment is required in comparison with the diffusion process.

With diffusion of dopants from the surface the density of doping atoms tends to be highest at the surface. One way of obtaining a surface layer with few doping atoms and hence of high resistivity is by epitaxial growth. If conditions are just right, atoms arriving at a crystal surface can fit onto the lattice and continue the growth of the crystal (that is what the term epitaxy implies). When the arriving atoms are the same as the bulk semiconductor with no impurities present, then a high-resistivity layer is produced. The growth process is slow, so only thin ($10 \, \mu m$) layers are made in this way.

Collisions

In passing through the crystal lattice, electrons and holes suffer collisions. We study the effect of collisions because they control the diffusion and drift of carriers, and hence the flow of electric current.

The first point to note is that carriers do *not* collide with the lattice atoms. The allowed wavefunctions have already taken account of the periodic lattice, in a fashion which predicts a plane travelling wave if the periodicity of the lattice is perfect.

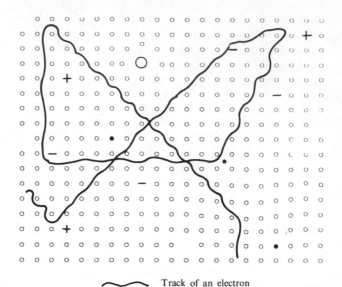

Track of an electron

Fig. 1.10.   This picture is intended to suggest the motion of an electron through a crystal lattice. The wavefunction of the electron fits the regular lattice, so the electron is not scattered from the lattice atoms, but may be scattered from some irregularity. In the picture there are 7 ionized impurities, 2 vacancies, 1 neutral impurity, 1 interstitial atom, and a phonon that can best be seen by viewing the picture from the side.

Abrupt changes of velocity occur when the lattice is imperfect, and in doped semiconductors two causes of irregularity— phonons and ionized impurities—cannot be avoided.

Phonons are lattice vibrations seen from a quantum point of view. The density of phonons in a solid is found to be proportional to the temperature T. The number of collisions an electron makes in a second is proportional to the distance it travels in a second, and to the density of scatterers, so that the collision frequency for electrons with phonons varies as (thermal velocity) $\times$ (phonon density), or as $T^{\frac{1}{2}} \times T$. Consequently the mean time between collision τ_d varies according to

$$\tau_d \propto T^{-\frac{3}{2}} \tag{1.17}$$

Scattering from phonons is the most important kind of collision in lightly doped semiconductors at room temperatures.

The frequency of scattering from donor and acceptor ions is

proportional to their concentration and is more effective when T is small. Thus ionized-impurity scattering dominates at low T, or in highly doped semiconductors.

Figure 1.10 shows some of the causes of scattering in a pictorial way, and also attempts to suggest the track of an electron.

Causes of current

Three reasons for a net flow of holes or electrons are
(a) an electric potential gradient dV/dx,
(b) a particle-number density gradient dn/dx,
(c) a temperature gradient dT/dx,
The last reason is discussed in Chapter 2, as it is important in such useful devices as thermoelectric cooling systems and power generators.

Drift current and mobility

If an electric field E accelerates carriers of charge q and effective mass m^* their acceleration f is given by

$$f = qE/m^*.$$

The distance each carrier is moved in the direction of E by this acceleration is $\frac{1}{2}f\tau^2$, where τ is the time between collisions. The average drift velocity v_d is found by taking an average value of the distance, and dividing by the average value of τ, thus

$$v_d = \frac{q}{m^*} \times \tfrac{1}{2}(\overline{\tau^2}/\bar{\tau}) \times E.$$

We may replace $\frac{1}{2}(\overline{\tau^2}/\bar{\tau})$ by a single symbol τ_d, the mean time between collisions as appropriate to carrier drift, then

$$v_d = \frac{q\tau_d}{m^*} \cdot E. \tag{1.18}$$

If we define $q\tau_d/m^* = \mu$, known as the *mobility* for the particular carriers in the material, then

$$v_d = \mu E.$$

Perhaps the easiest way of thinking of μ is merely as the constant relating v_d and E. Another way of putting the same idea is that μ is the velocity for unit field.

Notice that, for high μ, we need τ_d high and m^* small.

Consider the drift and current produced by the same field on holes and on electrons. The flux of particles is a useful concept—it is merely the flow of particles without regard for their charge.

	Electric field	Flux	Electric current
For holes	⟹	⟹	⟹
For electrons	⟹	⟸	⟹

The drift currents carried by both holes and electrons add to give a larger total if both are present in the same material. The current density j carried by drifting carriers is

$$j = nqv_d,$$

and if there are different kinds of carrier present they contribute separately to the current density.

The conductivity σ of a sample may be defined as the current that flows across unit cube when the field and hence voltage between opposite faces is unity. Thus for a semiconductor where both holes and electrons are present, if we define e to be $+1 \cdot 6 \times 10^{-19}\,\text{C}$,

$$\sigma = ne\mu_e + pe\mu_h. \tag{1.19}$$

Mobility is a useful generalizing concept when discussing the effect of electric field on carriers. A very useful first approximation is to say that μ is a constant for a given carrier and material. In fact μ depends on temperature, and on doping density when this is high, because of the effect on the collision time. Figure 1.11 shows these effects. The mobility also depends on the electric field: it falls at high fields, so that the drift velocity in many materials tends to a maximum limit. For silicon the maximum drift velocity is $10^5\,\text{m s}^{-1}$ for both holes and electrons. The field needed for electrons to reach this speed is about $2 \times 10^6\,\text{V m}^{-1}$, but holes require fields above $10^7\,\text{V m}^{-1}$, where avalanche ionization is beginning to set in. Nevertheless we shall work with a constant mobility for nearly all devices.

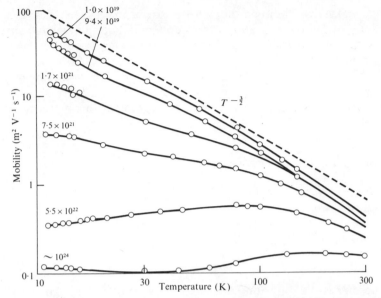

Fig. 1.11.   Mobility versus temperature for a series of samples of Ge doped with arsenic. The dashed line represents the theoretical slope for scattering by phonons. The number by each curve is the net doping density. (The data is taken with permission from E. M. Conwell (1952). *Inst. Radio Engrs. Proc.* **40,** 1327–1337.)

Diffusion of carriers

Diffusion can be looked at in two ways, (a) the independent-particle picture, and (b) the collective-average description.

The independent-particle picture

Here we think of each particle going on its way without regard to the presence or motion of the other diffusing particles; usually this is a valid assumption.

The motion can be described as a series of steps, of a fixed average length l, each of which is in a direction unrelated to the previous step. The style of progress has become known as 'the drunkard's walk'. Einstein showed that, on average, after N steps a particle is a distance $\sqrt{N} \times l$ away from its starting position (actually $\sqrt{(3N)} \times l$ in a three-dimensional situation). Using the simple $\sqrt{N} \times l$ formula, if a particle has to travel $10l$ then it will take 100 steps. The time taken for a carrier to cross the base of a transistor can be estimated if the step length and thermal velocity are known—now we can see why high-frequency transistors have narrow bases.

The net flow of particles down a density gradient occurs merely because there are more starting from the high-density regions: there is no force exerted by the density gradient forcing particles to diffuse.

The collective average description

This view is not in conflict with the independent-particle picture and predicted results should agree.

The flux F of particles down a density gradient dn/dx is described by

$$F = -D \, dn/dx, \qquad (1.20)$$

where D is the diffusion coefficient. Equation (1.20) is known as Fick's law, and applies to any example of diffusion. The electric-current density is

$$J = -qD \, dn/dx,$$

where q is the charge on the carrier.

If we consider the diffusion of holes and electrons in the same density gradient, we see that the hole and electron fluxes are in the same direction, and that the conventional electric currents thus tend to cancel—the opposite of the situation when drift was examined.

	Density gradient	Flux	Electric current
For holes	→	←	←
For electrons	→	←	→

D and μ are related by

$$D/\mu = \kappa T/e,$$

another relation due to Einstein; this relation is derived on p. 62. The resulting is not surprising, as D and μ are each an effect of collisions of carriers with the same lattice defects. Consequently a formula for D is

$$D = \kappa T \tau_d / m^*. \tag{1.21}$$

Generation and recombination of holes and electrons

This is a non-rigorous section, which shows how some ideas can be used to lead quickly to results.

The general principle

The transition rate between two states is the number of transitions per second from one state to the other. In equilibrium, transitions still occur, but the rate one way equals the rate the other way.

For a particular transition the rate is given by the formula

$$K = (\text{number of candidates}) \times (\text{number of places})$$

$$\times \begin{cases} \exp(-\Delta W / \kappa T) & \text{for an upward transition} \\ 1 & \text{for a downward transition,} \end{cases} \tag{1.22}$$

where K is a constant. For the upward and downward transitions between two states, the Ks are the same.

In a semiconductor in the conduction band, the number of places $= A_c - n \sim A_c$, and the number of candidates $= n$. In the valence band, the number of places $= p$ and the number of candidates $= A_v - p \sim A_v$. In this case we think of electrons in the valence band not holes.

Example 1.1. Equilibrium in an intrinsic semiconductor. We may equate the upward and downward transition rates. (The arrows indicate transitions to higher or lower energies).

$$\uparrow A_v A_c \exp\{-(W_c - W_v)/\kappa T\} = \downarrow pn = n_i^2.$$

Example 1.2. Equilibrium in a doped semiconductor (see Fig. 1.12). D is the number of donors and αD is the number of donors not ionized. Consider the equilibrium between donors and the conduction band where W_d is the small energy required to ionize a donor:

$$\uparrow \alpha D A_c \exp(-W_d / \kappa T) = \downarrow n(1 - \alpha)D;$$

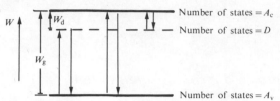

Fig. 1.12. Transitions between states in a doped semiconductor. The conduction band and valence band are represented by the equivalent densities of states A_c and A_v. Three pairs of transitions are possible.

hence

$$\frac{\alpha}{1 - \alpha} = \frac{n}{A_c} \exp(+W_d/\kappa T),$$

n/A_c is very small, and $\exp(+W_d/\kappa T)$ is not very large, hence $\alpha/(1 - \alpha)$ is small and α is small.

Donors are therefore nearly all ionized—as assumed earlier.

Example 1.3. Equilibrium between the conduction band and the valence band.

$$\uparrow A_v A_c \exp(-W_g/\kappa T) = \downarrow np, \tag{1.23}$$

as before. This is the semiconductor equation again.

Example 1.4. Equilibrium between the valence band and donors.

$$\uparrow (1 - \alpha)DA, \exp\{-(W_v - W_d)/\kappa T\} = \downarrow \alpha Dp. \tag{1.24}$$

Compare (1.23) and (1.24). Each side of (1.24) has a small term replacing a large one in (1.23), so the transition rates to the valence band are smaller than between the conduction band and the valence band if the Ks are the same.

Example 1.5. Minority-carrier recombination. When the carrier concentrations in a semiconductor are disturbed, and the cause of the disturbance is removed, then the concentrations return to normal. We can analyse this process.

Take n-type material to which equal numbers of excess holes and electrons have been added (how?). If equilibrium concentrations are n_0 and p_0 and excess concentrations are n_1 and p_1

$\uparrow$ (transition rate) $= KA_vA_c \exp(-W_g/\kappa T) = Kn_0p_0$,

as before, and

$\downarrow$ (transition rate) $= K(n_0 + n_1)(p_0 + p_1)$.

The rate of combination

$\downarrow - \uparrow = K\{(n_0 + n_1)(p_0 + p_1) - n_0p_0\}$,

but in n-type material $n_0 \gg n_1$, so the net rate becomes

$$-\mathrm{d}p_1/\mathrm{d}t = Kn_0p_1, \tag{1.25}$$

i.e. the excess minority carriers decay away exponentially with a lifetime τ_r which is given by $1/Kn_0$. In practice τ_r has to be determined by experiment as recombination can occur in several ways, for instance via traps in the middle of the band gap.

Au in Si forms traps, and the resulting value of τ_r is given by

$$\tau_r(s) = 2 \times 10^{10}/N_{Au} \ (\mathrm{m}^{-3}),$$

where N_{Au} is the concentration of the Au atoms. Thus if N_{Au} is $10^{19} \ \mathrm{m}^{-3}$, $\tau_r = 2 \times 10^{-9} \ \mathrm{s}$.

The gold also acts as a donor, and will cancel the effect of any acceptors; so in p-type material we must arrange that $N_{Au} \ll N_a$. Thus we cannot have both light doping and a very short lifetime in Si.

The average distance that a carrier can diffuse before recombining is known as the *diffusion distance L*. If the mean distance between collisions is λ, then in a time τ_r, a carrier can make τ_r/τ_d collisions, and on the independent-particle picture can diffuse a distance $\lambda(\tau_r/\tau_d)^{\frac{1}{2}}$. If the average thermal velocity is $(\kappa T/m^*)^{\frac{1}{2}}$, then eqn (1.21) can be used to show that

$$\lambda(\tau_r/\tau_d)^{\frac{1}{2}} = (D\tau_r)^{\frac{1}{2}} = L.$$

The continuity equation

The ideas on diffusion, drift, and recombination can be combined into a single equation, the continuity equation. This equation makes a useful starting point for further analysis. The equation expresses the idea that the difference between the

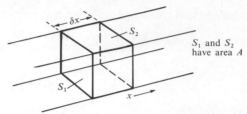

Fig. 1.13.　Charges enter the element of volume through S_1 and leave through S_2; any difference between the rate of entering and of leaving results in a change in the number contained in the volume.

numbers of particles entering a region and leaving it (no matter how) is equal to the change in the number of particles inside the region.

In Figure 1.13 the number of electrons entering per second by drift and diffusion across the surface S_1, of area A, is

$$-A\mu(nE)_x - AD(\mathrm{d}n/\mathrm{d}x)_x.$$

(We have taken a simple case where the electric field E and the density gradient $\nabla \cdot n$ have non-zero components in the x-direction only. The subscript x implies that the quantity is evaluated at x.)

The number of particles leaving at the surface S_2 is

$$A\mu(nE)_{x+\delta x} + AD(\mathrm{d}n/\mathrm{d}x)_{x+\delta x}.$$

The generation and recombination rates per unit volume are g and r, so the continuity equation is

$$A\mu(nE)_{x+\delta x} - A\mu(nE)_x + AD(\mathrm{d}n/\mathrm{d}x)_{x+\delta x} - AD(\mathrm{d}n/\mathrm{d}x)_x$$

$$+(g+r)A\,\mathrm{d}x = \frac{\mathrm{d}}{\mathrm{d}t}(nA\,\delta x).$$

When each term is divided by $A\,\delta x$, the first four terms may be recognized as differentials as $\delta x \to 0$, so the continuity equation can be written

$$\mu\frac{\mathrm{d}}{\mathrm{d}x}(nE) + D\frac{\mathrm{d}^2 n}{\mathrm{d}x^2} + g - r = \frac{\mathrm{d}n}{\mathrm{d}t}. \tag{1.26}$$

This is now a differential equation, and boundary conditions and forms for $E, g,$ and r need to be specified before further progress can be made. In later chapters some devices will be analysed—we shall take fairly simple situations, where some terms can be neglected and the rest are simple.

Equivalent circuits

Of all the general concepts used in electronics, probably only the idea of an equivalent circuit is both applicable to this book and sufficiently difficult to require some discussion. An equivalent circuit provides a representation of some aspect of the electrical behaviour of a device in a form suitable for inclusion in circuit calculations. Different equivalent circuits are needed for describing the way a device responds to small sine-wave signals or to large pulses, for its noise properties or for its d.c. characteristics.

Two common components of equivalent circuits are current and voltage sources (or generators). The circuit function of a source is defined by the symbols beside the diagram. Figure 1.14 shows some examples of such sources: a current generator which supplies a fixed current i_{CB_0}, flowing independently of any opposing voltage or other circuit parameter (Fig. 1.14(a)); a current generator supplying a current βi_B dependent on the value of another current i_B (β is a dimensionless constant) (Fig. 1.14(b)); a voltage generator, capable of supplying any current,

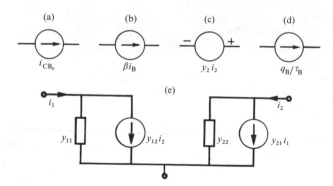

Fig. 1.14. Equivalent circuits: (a) a fixed-current generator or current source; (b) a dependent-current generator; (c) a dependent-voltage generator; (d) a current generator dependent on the charge q_B on another circuit element; (e) the y-parameter equivalent circuit for a three-terminal device.

controlled by a current i_2 via a susceptance y_2 (Fig. 1.14(c)); a current generator controlled by a charge q_B (Fig. 1.14(d)).

We very often require an equivalent circuit to describe the small-signal properties of devices. Two approaches are possible here—the formal and the physical.

Formal equivalent circuits

We can describe the way a device acts on signals in a circuit by four complex numbers which refer to just one signal frequency. Figure 1.14(e) shows one way of doing this. y_{11}, y_{12}, y_{22}, and y_{21} are the four complex numbers; they all have the dimensions of (current/voltage), and as a set are known as the 'y parameters' of the device. If instead of current generators we use voltage generators the result is a set of 'z parameters', while 'h parameters' include both current and voltage generators. These formal equivalent circuits represent exactly the behaviour of a device at a single frequency. Hence the variation with frequency of each parameter has to be specified (manufacturers on occasion supply graphs of this information). We can infer very little about the physical processes taking place in the device from the formal equivalent circuit.

Physical equivalent circuits

A physical equivalent circuit is based on what is happening in the device, and if skillfully constructed, suggests the physics while, at the same time, representing the behaviour of the device over a wide range of frequency. The description may omit some processes in the interest of simplicity, and hence will not be perfectly accurate. However, extra components can be added to a physical equivalent circuit to model any physical processes deemed worthy of inclusion, and thus (we hope) represent the behaviour more accurately.

A good physical model should therefore be flexible, allowing complexity to be traded for accuracy, and should be helpful in relating features of behaviour to the underlying physical principles.

PROBLEMS

1.1. Draw the graphs of $S(W)$, $P(W)$, and $N(W)$ for a p-type semiconductor, showing the Fermi level and the energy level of the acceptors.

1.2. Show that the curve for $N(W)$ has a maximum $\kappa T/e$ above the edge of the conduction band, so long as the doping is not too heavy. Interpret this physically.

1.3. Find the probability that a state is occupied by electrons at a temperature T, when the state is an energy W higher than the Fermi level, for the following cases: (a) $W = 0\cdot01$ eV and $T = 300$ K; (b) $W = 0\cdot6$ eV and $T = 300$ K; (c) $W = 0\cdot01$ eV and $T = 4$ K. Suggest situations where each calculation might be relevant.

1.4. Calculate the mean free path for electrons and for holes in Si from the mobility. Take the carrier temperature to be 300 K, and express the answer in terms of the distance between Si atoms $(2\cdot34 \times 10^{-10}\,\text{m})$.

1.5. Compounds of the form AB, where A is a group III element and B is a group V element, are semiconductors. By referring to the periodic table, list all such compounds.

1.6. Set up a computer simulation for carrier diffusion. Divide the distance through which carriers move into equal parts (ten is satisfactory). The diffusion process is modelled by arranging that the carriers in region N at time T move in equal numbers to region $(N+1)$ and to region $(N-1)$ at time $(T+1)$. At the boundaries of the whole distance carriers are injected or absorbed as appropriate for the physical situation being modelled. The diffusion of carriers through the base of a transistor is an instructive topic, and thought is required to relate computed times and distances to physical times and distances.

1.7. What is the velocity of electrons in Si when 1 V is applied across 1 mm? What is the velocity when 5 V are applied across 1 μm? What is the transit time in each case?

1.8. Gold doping concentrations range from 10^{20} to $10^{24}\,\text{m}^{-3}$. How far can carriers travel before recombining? Consider drift and diffusion in each case.

2. Bulk effects

Although most semiconductor devices exploit the properties of interfaces, there are a number of uses to which a uniform block of semiconductor can be put, and these form the subject matter of this chapter. The physics of a material is also conveniently studied in samples whose properties are uniform throughout. Chapters 3, 4, and 5 build on the concepts introduced in this chapter, examining the physics and applications of interfaces.

Resistance

The conductivity of a block of semiconductor is (eqn (1.19))

$$\sigma = ne\mu_e + pe\mu_h.$$

In semiconductors the conductivity can range between wide limits. The conductivity is a minimum and the resistivity a maximum when the semiconductor is intrinsic, that is when $n = p = n_i$. (*Exercise.* This is strictly true only when $\mu_e = \mu_h$. Work out the condition for minimum conductivity when $\mu_e \neq \mu_h$.)

For Ge, the maximum resistivity at 300 K is about $0.49 \, \Omega$ m. The equivalent figure for Si would be several thousand ohm metres, but it has not been possible to obtain pure enough Si, so that in practice several hundred ohm metres is the maximum resistivity.

The maximum conductivity is set by there being a limit to the solubility of doping atoms in the semiconductor lattice, though scattering from ionized doping atoms reduces mobility and the benefit of high doping. Resistivity in Ge and Si can be reduced below $10^{-5} \, \Omega$ m.

In integrated circuit processing, thin layers are used where the doping varies with depth below the surface. Consequently it is more convenient to work with the surface resistivity S, rather than the varying volume resistivity. S is quoted in ohms per square, as a square of any size has the same resistance between two opposite sides. Values of S in practice might range between 5 Ω per square and 500 Ω per square and it is found convenient to make resistors between 10 Ω and 10^5 Ω. Outside this range, other circuit techniques are used.

The resistance of a block of semiconductor varies with temperature. The effect is put to good use in some devices, but is usually not wanted. First we shall examine the dependence of n and p on temperature, then combine this information with the temperature variation of the mobility to give the dependence of σ (or ρ) on the temperature T. Figure 2.1 sketches the variation

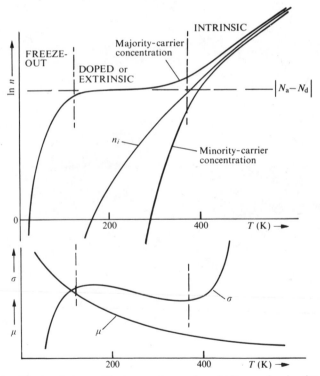

Fig. 2.1. The variation with temperature of majority- and minority-carrier concentration, and of mobility μ and conductivity σ for a semiconductor.

of majority and minority carrier density for a doped semiconductor. The intrinsic-carrier density increases steadily as T increases. The majority-carrier density is roughly constant over a fair range, and is about equal to the density of the uncompensated doping atoms. At high temperatures, the intrinsic density surpasses the doping density, and both majority and minority densities increase rapidly. At very low temperatures, the doping atoms are no longer ionized, and the majority-carrier density is low. If the majority-carrier density is required to be constant in a device, then the range of temperature marked 'extrinsic' in Fig. 2.1 is most satisfactory. Notice that the minority-carrier density depends strongly on temperature over all temperature ranges.

We can examine the rate at which n_i and the minority-carrier density vary with temperature by finding an expression for n_i. Multiply eqns (1.10) and (1.11), and take the case when $p = n = n_i$:

$$n_i = \frac{2}{h^3} (2\pi \kappa T)^{\frac{3}{2}} (m_e/m_h)^{\frac{3}{4}} \exp\{-(W_c - W_v)/2\kappa T\}. \quad (2.1)$$

Differentiating by parts with respect to T, and then simplifying gives

$$\frac{dn_i}{dT} = \frac{n_i}{T} \left(\frac{3}{2} + \frac{W_c - W_v}{2\kappa T} \right)$$

where $W_c - W_v$ is the band-gap energy W_g, so we may write

$$\frac{dn_i}{n_i} = \frac{dT}{T} \left(\frac{3}{2} + \frac{W_g}{2\kappa T} \right).$$

For Si, W_g is $1.1\,\text{eV}$ and κT is about $25\,\text{meV}$ at $300\,\text{K}$, so that $W_g/2\kappa T \sim 22$, and $dn_i/n_i = (dT/T) \times 23\frac{1}{2}$. At $300\,\text{K}$, if dT is $1\,\text{K}$, then dn_i/n_i is $23\frac{1}{2}/300$ or about 8 per cent, so that n_i in Si at $300\,\text{K}$ increases by 8 per cent for a $1\,\text{K}$ temperature rise. The minority-carrier density is equal to $n_i^2/(A - D)$, so it will vary by 16 per cent for $1\,\text{K}$ temperature change. In Ge W_g is smaller, and the minority-carrier density only varies by 10 per cent for $dT = 1\,\text{K}$.

If we consider semiconductors that are not too heavily doped, at temperatures around $300\,\text{K}$ (this covers a lot of cases), then mobility (μ) varies roughly as $T^{-\frac{3}{2}}$. The variation of μ is only detectable as a change in σ when the majority-carrier density is not changing. Over the 'extrinsic' range of temperature (see Fig. 2.1), the resistance of a semiconductor increases with temperature, because μ is falling. Nevertheless, the common statement that the resistance of semiconductors falls as temperature rises is not wrong, if we interpret it as applying only to intrinsic semiconductors.

Thermistors are semiconductor resistors whose temperature dependence is exploited; they are available with either a positive or negative temperature coefficient of resistance. A thermistor can be used as a transducer to provide an electrical signal describing the temperature of its surroundings, or as a non-linear circuit element, where the resistance changes as the electrical power dissipated in the thermistor is varied. In either use, the thermistor responds only to slow changes, perhaps below 1 Hz.

Growth of 3–5 materials

Most devices made of 3–5 semiconductors have a number of thin layers of differing composition on a thicker substrate. The substrate will be a slice of a binary 3–5 compound (GaAs, GaP or InP) cut from a single-crystal boule 2 or 3 in. in diameter. The boules are grown in similar ways to those used for silicon. Several techniques have been developed which allow ternary and quaternary epitaxial layers to be grown on binary substrates.

Liquid phase epitaxy (LPE)

A substrate slice is held in a graphite container, and is slid into contact with liquid from which the epitaxial layer can crystallize out. Thus to deposit a layer of GaAlAs on GaAs, the liquid would be GaAs and Al dissolved in molten Ga, at about 750°C. The temperature is adjusted so that the substrate grows rather than dissolves.

Growth rates are around 1 μm min^{-1} and thicknesses of 1 to 10 μm are usual. Up to five different layers can be grown, one on top of another, by sliding the slice from under one reservoir of liquid to another. This method was the first to be widely used, and is the way semiconductor lasers and LEDs are made.

Vapour phase epitaxy (VPE), Chemical vapour-phase deposition (CVD), Metallo-organic vapour-phase deposition (MO–CVD)

The atoms to form the new growth are carried as volatile chemicals in a gas stream (usually H_2) over the substrate. Total gas pressure may be 0·1–1 atm. When the molecules meet the heated substrate, they decompose, leaving the group 3 or group 5 atoms to build up the epitaxial layer. The chemicals that are used include:

AsH_3	$AsCl_3$	$P(CH_3)$	trimethyl phosphide TMP
PH_3	PCl_3	$Al(CH_3)$	trimethyl aluminium TMA
H_2S		$Ga(CH_3)$	trimethyl gallium TMG
H_2Se		$Cr(CO)_6$	

Where compounds of metal with organic radicals are used, the name 'metallo-organic chemical vapour-phase deposition' is used. The composition of a growing layer can be varied by changing the composition of the gas flowing over the substrate, so that it is possible to grow a hundred of more layers on a single substrate just by turning gas flow taps on and off.

Growth rates are less than for LPE, about 0·1 to 0·5 μm min^{-1}. Since 1 μm is about 3000 atomic layers, this is a rate of about 20 to 100 atomic layers per second.

Molecular Beam Epitaxy (MBE)

The atoms to form the new growth are evaporated in high vacuum from separate heated containers, so that a beam of molecules of each required element falls on the substrate. The rate of arrival of each kind of atom is regulated by the temperature of its container, and by a cover that can be opened or shut. The atoms only come together on the new crystal surface, and each is separately controlled, so this method is more flexible than LPE or CVD. However the growth rate is low (0·01–0·05 μm per minute) and the equipment is elaborate and expensive. Multiple layers of binary, ternary or quaternary compounds can be deposited, and doping atoms introduced as required.

Deep levels and semi-insulating semiconductors

An impurity that produces donor states towards the centre of the forbidden band (Fig. 2.2) has the effect of reducing the free carrier concentration towards the intrinsic value. Very few deep level donors will provide electrons in the conduction band, as the energy required is too great. The deep level donors can however provide electrons to fill any lesser number of normal acceptors, as no energy is needed for this. The acceptors are now unable to accept electrons from the valence band, so there are very few holes in the valence band.

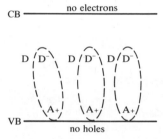

Fig. 2.2. Deep donors outnumber normal acceptors.

The result is that there are very few holes or electrons free to carry current. This kind of material is known as semi-insulating (SI), and SI GaAs doped with Cr is often used as a substrate for GaAs epitaxial layers. Devices formed in the epitaxial layers will be coupled to a less extent via the semi-insulating substrate than they would be if the substrate had a high conductivity. If the deep donors outnumber the shallow acceptors, then the deep donors act as traps, and are discussed in Chapter 3. (*Exercise*— check that deep acceptor states would produce SI material.)

Thermoelectricity

When temperature differences exist between parts of a conducting system, thermoelectric effects will occur. These are stronger in semiconductors than in metals, and can be a cause of incorrect circuit operation, or can be exploited for useful

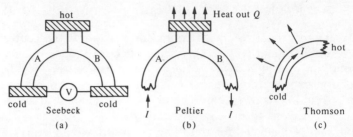

Fig. 2.3. There are three ways in which thermoelectric effects can be demonstrated: (a) the Seebeck effect—a temperature difference across two materials causes an output voltage; (b) the Peltier effect—a current causes heat flow into or out of a junction; (c) the Thomson effect—a current along a temperature gradient causes heat flow into or out of the conductor.

functions. There are three thermoelectric effects—Seebeck, Peltier, and Thomson—all are thermodynamically reversible (Fig. 2.3).

The Seebeck effect is the thermocouple effect. When the junctions of two different conductors are at different temperatures an e.m.f. is produced. For two materials A and B, the Seebeck coefficient S_{AB} is defined as

$$S_{AB} = \mathrm{d}\,V_{AB}/\mathrm{d}T.$$

The Peltier effect is used in thermoelectric power generating and cooling. When a current I is passed from one material to another, heat Q is absorbed (or liberated) at the junction. The Peltier coefficient is

$$\Pi_{AB} = \frac{Q}{I_{AB}}.$$

The Thomson effect is of lesser importance numerically. When a current I flows in a conductor between two regions at different temperatures, heat is released (or absorbed) through the volume of the conductor. The Thomson coefficient is

$$\tau = Q/I\Delta T,$$

where ΔT is the temperature difference between the two parts of the conductor and Q is the whole of the Thomson heat released.

The Thomson heat should not be confused with Joule heating—the familiar I^2R power loss in a resistor which is not reversible.

A thermodynamic analysis shows that there are relations among the thermoelectric coefficients.

$$\Pi_{AB} = TS_{AB},$$
$$\tau_A = T\, dS_A/dT.$$

Here S_A means the part of S which depends on material A.

Since the three coefficients are related, only one needs to be analysed in detail. The Seebeck coefficient for an n-type semiconductor will be analysed. The major contribution to the Seebeck coefficient comes from the change in Fermi level with temperature. From eqn (1.10) the difference in energy between conduction band and Fermi level is

$$W_C - W_F = \kappa T \ln\left\{\frac{2(2\pi m_e \kappa T)^{\frac{3}{2}}}{h^3 n}\right\}.$$

Therefore, the first contribution S_F to S will be

$$S_F = \frac{1}{e}\frac{d(W_C - W_F)}{dT}$$
$$= \frac{W_C - W_F}{eT} + \frac{\kappa T}{e}\left\{\frac{3}{2T} - \frac{d}{dT}\ln(n)\right\}.$$

The term in $d\{\ln(n)\}/dT$ cancels out eventually, so does not need to be evaluated. The remaining contributions to S are obtained from the balance between drift and diffusion on open circuit.

In an n-type semiconductor

$$j = ne\mu \frac{dV}{dx} - \frac{d}{dx}(Den) = 0. \tag{2.2}$$

Usually dn/dx is taken to non-zero and dD/dx is zero, but in this topic the change of D with temperature is also considered.

First, assume D varies while n is constant. Then

$$ne\mu \frac{dV}{dx} = \frac{dD}{dx}\cdot en$$

or

$$dV/dx = \frac{1}{\mu} \cdot dD/dx.$$

The Seebeck coefficient S_D due to change of D is

$$S_D = \frac{dV}{dT} = \frac{dV}{dx} \cdot \frac{dx}{dT} = \frac{1}{\mu} \frac{dD}{dx} \cdot \frac{dx}{dT}$$

$$= \frac{1}{\mu} \frac{dD}{dT}.$$

The diffusion coefficient D changes with temperature differently in different materials, but a law

$$D(T) = D_0 T^{\frac{1}{2}}$$

is not uncommon. In that case

$$S_D = \frac{kT}{e} \cdot \frac{1}{D_0 T^{\frac{1}{2}}} \frac{d(D_0 T^{\frac{1}{2}})}{dT} = \frac{k}{e} \frac{T}{2T} = \frac{k}{2e}.$$

The third contribution S_n to S comes from dn/dT. We start again with eqn (2.2), assuming D is constant

$$ne\mu \frac{dV}{dx} = De \frac{dn}{dx}.$$

Then

$$S_n = \frac{dV}{dT} = \frac{dV}{dx} \frac{dx}{dT} = \frac{D}{\mu} \frac{1}{n} \frac{dn}{dx} \cdot \frac{dx}{dT}$$

or

$$S_n = \frac{\kappa T}{e} \cdot \frac{1}{n} \frac{dn}{dT} = \frac{\kappa T}{e} \frac{d}{dT} (\ln(n)).$$

This cancels a term in S_F as noted earlier. The full value of S is now

$$S = S_F + D_D + S_n = \frac{\kappa}{e} \left[2 + \frac{W_C - W_F}{\kappa T} \right].$$

For heavily-doped semiconductors, $(W_C - W_F)/\kappa t$ is in the range 10 to 30 and is the most important term. For p-type semi-conductors, S has the opposite sign and a similar magnitude. Metals have values of S roughly equal to κ/e.

The commonest occasion where a thermoelectric effect is apparent is in a hot p–n junction in an electronic device.

The Seebeck voltage is $S\Delta T$, so that if we take $S = 30\,\kappa/e$

$$S\Delta T = \frac{30}{T} \cdot \frac{\kappa T}{e} \cdot \Delta T \simeq 2 \times 10^{-3}\,\Delta T.$$

This means that a voltage change of $2\,\mathrm{mV}$ is generated at a p–n junction each time the temperature changes by $1\,\mathrm{K}$. Such voltages are a major cause of drift in high-gain d.c. amplifiers. The designer's cure is to make the output from an amplifier depend on the difference of the signals from two devices which are at very similar temperatures.

Long thin conductors are used as temperature transducers—thermocouples. Metallic alloys are usually used for this purpose. Short fat conductors are used for the generation of electrical power and for refrigeration.

In Fig. 2.4, one pair of elements of a Peltier refrigerator is shown. A current I is passed through the two blocks of semiconductor, and heat Q_{cold} is absorbed at the cold junction and transferred to the hot junction. Heat flows in the opposite direction (hot-to-cold) by normal thermal conduction, and there is also I^2R heating in the semiconductors, of which half will flow to the cold junction. The heat balance can be expressed as

$$Q_{\text{cold}} = \Pi T - \tfrac{1}{2}I^2R + K\Delta T,$$

where K is the thermal conductance of the device and R is its

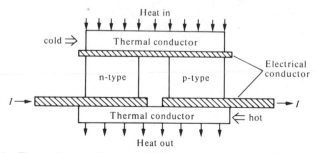

Fig. 2.4. For optimum efficiency, Peltier cooling units are short and squat. Currents are about $10\,\mathrm{A/cm^2}$.

resistance. Then

$$\Delta T = Q_{cold} - \Pi I + \tfrac{1}{2} I^2 R / K.$$

The maximum cooling is found when $\{d(\Delta T)/dI\} = 0$. If the thermal load Q_{cold} is zero, or in other words the cold junction is well insulated

$$\Delta T = -\frac{1}{2} \frac{\Pi^2}{RK} = -\frac{Z}{2}.$$

The term Z is a figure of merit for a Peltier cooling device, and it combines the Peltier effect, electrical resistance, and thermal conduction. The best materials are semiconductors with heavy atoms, as these have lower values of K. Alloys of Bi, Te, and Sb are used for Peltier cooling devices, and temperature reductions of 100 K can be achieved. Currents are high, and a number of junctions are connected in series to give more convenient operating conditions.

There is a quick way of checking whether a sample of semiconductor is n-type or p-type. The method uses the Seebeck effect, which is the basis of the thermocouple. To carry out the test, touch one hot and one cold lead from a voltmeter capable of responding to 10 mV onto the semiconductor sample. Where the hot lead touches, the semiconductor will become hot and form one arm of a thermocouple, the copper leads forming the other arm. If the hot arm is positive, the semiconductor is n-type.

The Hall effect

When a current of density **J** flows in a semiconductor perpendicular to a magnetic field **B** (Fig. 2.5), there is a force on the holes and electrons perpendicular to both **B** and **J**, and a transverse current tends to flow. The process is known as the *Hall effect*, and provides a way of estimating the majority-carrier density and sign, and is also the basis of devices for measuring magnetic field and for multiplying two signals.

In Fig. 2.5, if J_x and B_z are as shown, then, because holes and electrons have oppositely directed drift velocities and opposite signs of charge, both of them will tend to be accelerated in the direction of the positive y axis by a force $q v_x B_z$, where v_x is the

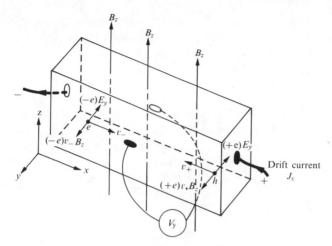

Fig. 2.5. The Hall effect in a semiconductor. Electrons and holes drift in opposite directions, but the transverse $(\pm e)v_{\pm}B_z$ force is in the same direction for both of them.

drift velocity of the carriers, and is given by

$$J_x = N_{ma}qv_x.$$

N_{ma} and q are the concentration and charge of the majority carriers (we are dealing with cases where the minority carriers are too few to be evident).

If the circuit is arranged so that the transverse current J_y is zero, the transverse force qv_xB_z must be balanced by a transverse electric force qE_y so we can write

$$qE_y = qv_xB_z = J_xB_z/N_{ma}. \tag{2.3}$$

Equation (2.3) is often rearranged

$$1/qN_{ma} = E_y/J_xB_z = R,$$

where R is called the Hall constant for the sample. It is negative for n-type semiconductors and for simple metals, and positive for p-type semiconductors, where the majority carriers are holes.

The three quantities J_x, B_z, and E_y may all be measured so that N_{ma} can be determined. A measurement of resistivity on the same sample gives μN_{ma}, and thus in combination with the Hall measurement yields a value for the mobility of the majority carriers.

The use of the Hall effect to measure magnetic field is straightforward. A sample has a fixed current passed through it, and is calibrated in a known magnetic field. InSb is a useful material for a magnetic field probe, as it has a high electron mobility.

If the magnetic field is produced by current flowing in a coil, eqn (2.3) shows that the output E_y is proportional to the product of the value of B_z, produced by the coil current, and J_x. Thus the output is a constant times the product of two currents—a result there are few other ways of achieving.

The transverse force on an electron ($-e\mathbf{v} \times \mathbf{B}$) tends to make it move in a circle. If it makes no collisions on the way, the angular frequency of rotation is given by

$$eB/m = \omega_c,$$

the electron cyclotron frequency. (Check this using $F = mv^2/r$ for motion in a circle.)

If τ_d is the electron collision frequency, the product $\omega_c\tau_d$ may be written, using the definition of μ following eqn (1.18),

$$\omega_c\tau_d = \frac{eB}{m^*}\tau_d = \frac{e\tau_d}{m^*}B = \mu B.$$

For a magnetic field to have a large effect on the electrical properties of a semiconductor, the circular motion of the electrons must be interrupted only seldom by collisions. Hence a simple criterion for a high magnetic field, in a semiconductor context, is

$$\text{either } \omega_c\tau_d \gg 1 \quad \text{or } \mu B \gg 1.$$

A semiconductor immersed in a strong magnetic field will absorb energy from a signal at the frequency ω_c. Since the only property of the semiconductor which affects the value of ω_c is the effective mass m^*, we now have a method of determining m^* by observing the frequency at which power is absorbed in a known magnetic field. The condition $\mu B \gg 1$ has to be satisfied.

If a magnetic field is parallel to the flow of current it can still have an effect on carrier motion. The effect is known as magnetoresistance—the resistance increases when the magnetic field is present.

Piezo-resistance

Strain gauges are used to measure the amount by which a body is distorted in some way—stretching, compressing, or twisting. The change in resistance of a silicon resistor when strained is known as piezoresistance, and is the physical effect used in silicon strain gauges.

Silicon and germanium have the same crystal structure as diamond and each unit cell is symmetrical about its centre. As a result silicon and germanium do not generate a voltage when they are strained and so are not piezo-electric substances. However they have large piezo-resistive coefficients in some directions and silicon is used to make sensitive strain gauges.

The fractional change of resistance $\Delta R/R$ for a given strain $\Delta L/L$ in n- or p-type silicon varies with direction, and reaches values that are 100 times greater than are found in metals. The reasons for this may be understood by looking back at Fig. 1.6. The resistance of a block of semiconductor is given by

$$R = \frac{\rho L}{A} = \frac{1}{\sigma}\frac{L}{A} = \frac{1}{ne\mu}\frac{L}{A}.$$

For a doped semiconductor, n is held at a fixed level by the density of doping atoms, but μ can change when the semiconductor is strained.

In n-type Si, the fundamental effect of a tensile strain is to depress the energy of those conduction-band minima that lie in the direction of the strain. Some of the electrons that populated these minima transfer to the other minima. Their effective mass for motion parallel to the strain is now lower, so the average mobility of the whole electron population rises. Figure 2.6 shows the effect for strain and resistance measured along three different directions in the crystal.

The strongest effect occurs in the (100)-direction, when one pair of minima are fully affected and the other four not affected at all. When the strain is applied along (111), the resistance changes very little, as symmetry requires, while strain along (110) has an intermediate effect.

If p-type silicon is used, the directions showing high or zero sensitivity are different from those in n-type silicon. This reflects the different symmetry of the valence-band structure, as

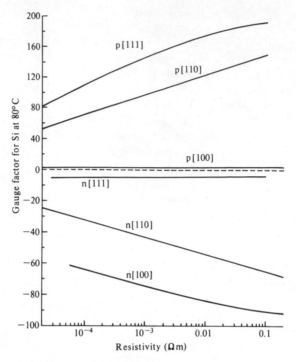

Fig. 2.6. The gauge factor is $(\Delta R/R)/(\Delta L/L)$. It is high for lightly doped silicon, and is between 1 and 2 for metals. (Taken by permission from Dean and Douglas (1962). *Semiconductor and conventional strain gauges.* Academic Press, New York.)

discussed in Chapter 1. Strain gauges are probably unique in requiring advanced concepts like tensor effective mass for describing the function of a simple device. Practical strain gauges are made as separate components, or are integrated with buffer amplifiers on a single silicon chip.

Photoconductivity

The basic process of photoconductivity is the absorption of photons by the semiconductor, resulting in the production of free carriers. Photoconductors exploit the resulting change of conductance. When the photon raises an electron from the valance band to the conduction band (Fig. 2.7(a)), producing both a free electron and a hole, we have intrinsic photoconduction. When

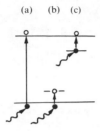

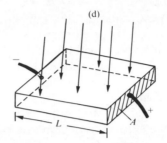

Fig. 2.7.   Photoconductivity: (a) intrinsic; (b) and (c) impurity photoconduction; (d) light falls on the photoconductor and produces C electron-hole pairs.

the photon ionizes a donor or acceptor, we have impurity photoconduction (Fig. 2.7(b)). In the latter case, only one free carrier is produced, and the energy required is much less than that of the band gap of the semiconductor.

The impurity semiconductors are used in an unusual state. They are cooled to a low enough temperature for the impurities not to be ionized, until such time as energy is provided by the arrival of photons. The ionization energy of Au acceptors in Ge is 0·15 eV, equivalent to 9 μm wavelength. Au-doped photoconductors are cooled with liquid nitrogen to ensure that the Au is not ionized. Zn-doped Ge (ionization energy 0·03 eV) is used to detect 40 μm far infra-red radiation, and is cooled to liquid helium temperatures. Thus the general rule is that the longer the wavelength of the radiation, the lower the temperature to which the detector has to be cooled.

In passing through an absorbing medium a beam of light is attenuated according to the law

$$-dn = n_0 \exp(-x/\lambda) \times (1/\lambda) \times dx = (1/\lambda)n(x)\,dx.$$

The chance of a photon being absorbed in a small distance dx is thus dx/λ. In intrinsic photoconductors, every atom can absorb, and λ may be $10^{-8} - 10^{-6}$ m (how many atomic diameters is this?). In impurity photoconductors, only the impurity atoms can absorb; there may be few of these, and λ may be much longer. A way of increasing the chance of a photon being absorbed is to surround the photoconductor with a reflecting cavity, so that photons pass and repass through the detector.

The carriers liberated by photons are used to carry current, so we are interested in how long they are available. In Fig. 2.6, a beam of light is producing C electron-hole pairs per second in a semiconductor. C will depend on the number of photons arriving at the surface, the fraction of these not reflected, and the fraction of those which are absorbed which produce electron-hole pairs rather than heat.

If the electron and hole lifetimes are τ_e and τ_h in the steady state, the extra carrier numbers (not densities) are

$$\Delta N = C\tau_e \quad \text{and} \quad \Delta P = C\tau_h.$$

The change in the conductance Δg equals $\Delta \sigma A/L$ where $\Delta \sigma$ is the change in the conductivity ($\Delta \sigma = e(\Delta N \mu_e + \Delta P \mu_h)/AL$), A is the cross-sectional area of the semiconductor, and L is its length. Thus

$$\Delta g = \frac{e}{L^2}(\Delta N \mu_e + \Delta P \mu_h) = \frac{eC}{L^2}(\mu_e \tau_e + \mu_h \tau_h). \tag{2.3}$$

For Δg to be large, $\mu_e \tau_e$ and/or $\mu_h \tau_h$ should be large and L should be small. The values of τ_e and τ_h can be adjusted over wide ranges by controlling trap densities, but when τ_e and τ_h are long the frequency response of the device may be poor. To see this consider C to be made up of a steady and sinusoidal part,

$$C = C_0 + C_1 \exp(j\omega t).$$

We ignore C_0 and the corresponding steady parts of ΔN and ΔP in this a.c. analysis. The rate of change of electron density is due to the varying part of C and to recombination, thus

$$\frac{d\Delta N}{dt} = \frac{-\Delta N}{\tau_e} + C_1 \exp(j\omega t).$$

If we try a solution $\Delta N = \Delta N_1 \exp(j\omega t)$, then

$$\Delta N_1 = C_1 \tau_e/(1 + j\omega \tau_e),$$

with a similar equation for holes. Therefore

$$\Delta g(\omega) = \frac{eC_1}{L^2}\left(\frac{\mu_e \tau_e}{1 + j\omega \tau_e} + \frac{\mu_h \tau_h}{1 + j\omega \tau_h}\right). \tag{2.4}$$

When $\omega\tau_{e,h} \gg 1$, Δg falls as $1/\omega$, so that $1/\tau_{e,h}$ represents a cut-off frequency for the device.

An implicit assumption is that carriers recombine before they drift to the end-contacts of the device. However, even if the carriers do reach the contacts, the equations remain unchanged. Consider two cases:

(a) One kind of carrier is trapped. Then as electrons drift out of the positive end of the semiconductor, more must enter at the negative end to keep the whole device neutral. Only when recombination at a trap occurs can ΔN or ΔP fall.

(b) Neither kind of carrier is trapped. Electrons will still enter at the negative contact and holes at the positive contact to replace the carriers that have drifted away. The excess carrier density can fall only by the recombination process. As a result, each photon may allow many carriers to pass through the photoconductor, especially if it is short. The process is known as photoelectric gain, and, as shown by eqn (2.4) is a demonstration of the way sensitivity may be traded for frequency response.

The Haynes–Shockley experiment

This experiment used the photoproduction of electron-hole pairs to demonstrate the importance of minority carriers, and provided a way of measuring their mobility, complementing Hall measurements which give the mobility of majority carriers.

Electron-hole pairs are produced by a pulsed light source focused onto a small area of semiconductor (Fig. 2.8). These carriers drift in opposite directions in an applied electric field.

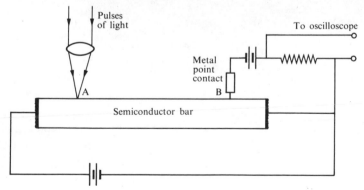

Fig. 2.8. The Haynes–Shockley experiment: Minority carriers are produced by pulses of light at A and drift to B where they can be detected.

The minority carriers are detected by a reverse-biased metal point contact (see p. 156 for an explanation of why a metal point can detect minority carriers). The arrival of the minority carriers is observed on an oscilloscope to be delayed by a time which is proportional to the distance the carriers have travelled, so the minority carriers have a definite drift velocity.

The drift velocity is found to be proportional to the drift field, so the minority-carrier mobility can be calculated. The value of the mobility for a specific carrier (e.g. electrons in Si) turns out to be the same whether they are majority or minority carriers, and the experiment shows that minority carriers are needed for a complete explanation of semiconductor phenomena.

Neutrality

Because semiconductors contain mobile electric charges, they tend to be electrically neutral, which is to say they contain equal amounts of positive and negative charge. It is interesting to see how large a region can be non-neutral, without there being large potential differences.

Let us consider p-type material with a region where a varying potential tends to a constant value, taken as zero. In the constant potential, the hole density p_0 will equal the acceptor density A. Where the potential has changed to V, the hole density will be controlled by the Maxwell–Boltzmann relation, as in eqn (1.9).

$$p = p_0 \exp(-eV/\kappa T).$$

Poisson's equation for this situation is

$$\frac{d^2V}{dx^2} = \frac{-e}{\varepsilon_r \varepsilon_0}(p - A) = \frac{-ep_0}{\varepsilon_r \varepsilon_0}\{\exp(-eV/\kappa T) - 1\}$$

where ε_0 is the permittivity of free space, and ε_r is the relative permittivity of the semiconductor.

This is unpleasant to solve in the general case, but when $|eV/\kappa T| \ll 1$, we can use the first two terms in a series approximation for the exponential, giving

$$\frac{d^2V}{dx^2} = \frac{e^2 p_0}{\varepsilon_r \varepsilon_0 \kappa T} V. \tag{2.5}$$

Equation (2.5) has a solution

$$V = V_0 \exp(-x/\lambda_D), \qquad (2.6)$$

where $\lambda_D = (\kappa T \varepsilon_r \varepsilon_0 / e^2 p_0)^{\frac{1}{2}}$ and is known as the Debye length. Equation (2.6) shows that a perturbation in the potential tends to build up or die away over distances of the order of λ_D. For a doping density of $10^{22}\,\text{m}^{-3}$ in Si at 300 K, $\lambda_D = 5 \times 10^{-8}\,\text{m}$.

Major field changes occur over distances greater than λ_D; the most important example of this happening is in the depletion layer in a p–n junction, which is examined on p. 62.

Semiconductor plasmas

Sometimes the electrical properties of a semiconductor cannot be explained by analysing the behaviour of a typical carrier and then multiplying by the number of carriers present. On occasion, it is necessary to consider all the carriers from the start of the analysis; the semiconductor is then said to be behaving as a plasma, and to be exhibiting collective properties. One such property, the tendency to neutrality, has been discussed in the previous section.

Another plasma effect is a collective oscillation at the plasma frequency ω_p, which equals $(ne^2/\varepsilon_r \varepsilon_0 m)^{\frac{1}{2}}$, where n is the concentration of mobile charges. Consider a region of space containing mobile electrons and an equal number of fixed positive charges. If the electrons are displaced bodily a distance x, as in Fig. 2.9, there will be a layer of positive charge on the

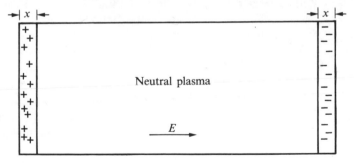

Fig. 2.9. The mobile electrons have all moved to the right leaving a layer of positive charge at the left.

left of the neutral plasma, and a layer of negative charge on the right.

In a charge layer of thickness x where there is a charge density en, the field changes by $enx/\varepsilon_r\varepsilon_0$. Inside the neutral region the field is uniform, so the force on a particle in this region is $e \times enx/\varepsilon_r\varepsilon_0$. The acceleration of the particle is then given by

$$\frac{\mathrm{d}^2x}{\mathrm{d}t^2} = \frac{-e^2n}{\varepsilon_r\varepsilon_0m} \cdot x,$$

which is the defining equation for simple harmonic motion with angular frequency $\omega = (ne^2/\varepsilon_r\varepsilon_0m)^{\frac{1}{2}}$, and represents the mobile charges performing a longitudinal oscillation.

In this analysis collisions have been ignored, and in semiconductors the collision frequency is often greater than ω_p. As a result, the effect of oscillations at the plasma frequency is seldom directly observable, but is more likely to be involved indirectly in a detailed analysis of the performance of a device. Nevertheless plasma effects can be directly observed in semiconductors; an example is the propagation of circularly polarized electromagnetic waves (helicon waves) in doped semiconductors in magnetic fields.

Gunn diodes

A uniformly doped sample of n-GaAs (or some other 3–5 semi-conductor) when provided with two end-contacts, can function as a microwave oscillator, and is then known as a Gunn diode. The functioning of a Gunn diode depends on the unusual drift velocity versus field curve shown in Fig. 2.10(a), which is a direct consequence of the details of the band structure of GaAs (Fig. 2.10(b)). There are minima in the conduction band at two energies, and electrons have very different effective masses and hence mobilities near the two minima. In the lower valley electrons are 'light' and have a high mobility. This is the region normally occupied. When the electrons acquire energy from a high electric field, they tend to be scattered into the higher valley, where their effective masses are much greater and their mobilities much reduced.

The current through a sample is proportional to the drift velocity, and the voltage across it to field, so Fig. 2.9(a) can be

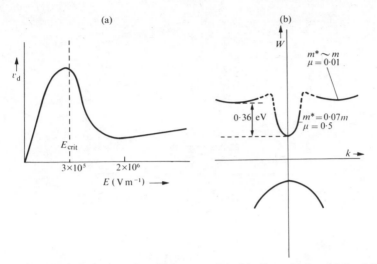

Fig. 2.10.   GaAs has properties which are exploited in Gunn diodes: (a) the drift velocity (v_d) falls sharply when the electric field (E) rises above E_{crit}: (b) electrons occupy the sharp lowest minimum of the conduction band when $E < E_{crit}$, but are scattered into the higher energy minima when $E > E_{crit}$.

regarded as an *I-V* characteristic for a sample. Notice that there is a region where the differential resistance is negative; this is a strong indication that unstable behaviour and oscillation is likely.

Instability manifests itself in two ways. The applied voltage is not distributed uniformly over the length of the device, and this non-uniform field changes as time goes by. Gunn diodes oscillate in a variety of ways, but we shall consider one mode only for further examination. In this mode a local high-field region forms near the negative electrode (the cathode), and propagates through the diode to the anode (Fig. 2.11). The double change of field is associated with a dipole-charge layer, a depletion region in front and an electron-rich region behind (check that this fits with Poisson's equation). The electron-rich layer is in the high field, so the electrons are 'heavy' and move no faster than the 'light' electrons in the low field. The dipole layer and local high field are known as a domain. The domain, having been formed at the negative contact, travels to the positive contact. On reaching it there is an extra pulse of current, the average field rises as there is no high field domain in the diode, and a new domain forms at the cathode to repeat the cycle.

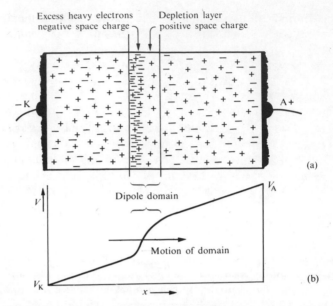

Fig. 2.11. Gunn diode domain: (a) A sketch of the dipole domain suggests the depletion layer at the head of the domain followed by the accumulation layer (the positive charges are fixed donor ions). (b) The voltage rises more rapidly in the domain than in the rest of the diode.

The domain travels with the velocity of the drifting electrons, so that the period of oscillation is l/v_d if the time required for domain formation is neglected (l is the length of the diode, and v_d is the electron drift velocity). The drift velocity is about 10^5 m s^{-1}, so that a Gunn diode 10 μm long should oscillate at 10 GHz.

If this mode of oscillation is to occur, the domain must be usefully smaller than the device. At the leading edge of the domain, the maximum space-charge density will occur if there are no electrons present, and will equal eN_d. If we integrate Poisson's equation across the depletion layer we get

$$\delta l e N_d = \varepsilon_r \varepsilon_0 (E_{max} - E_{min})$$

where δl is the width of the depletion layer, E_{max} is the high field behind the depletion layer, and E_{min} is the low field in the bulk in front of the depletion layer. E_{max} is about 2×10^5 V m^{-1}, while E_{min} is much less and the exact value is not important. The product $\delta l N_d = 10^{14} \, \text{m}^{-2}$, and thus sets a lower limit to the length of material which can set up a travelling domain and oscillate in this mode. To give the domain a chance to build up and propagate, $l > 10 \delta l$, so we have as a design criterion

$$l N_d > 10^{15} \, \text{m}^{-2}.$$

The voltage supplied to Gunn diodes has to form both the low-field and the high-field regions. Either may be the larger contribution, so a simple calculation of the voltage will not be attempted. Often a voltage supply between 5 V and 20 V is needed. The current taken by the diode is proportional to its area and can be chosen by the designer. With good heat sinks, continuous wave (c.w.) operation is possible, though pulsed operation is more usual. An example has

Power output	Frequency
250 mW c.w.	18 GHz

PROBLEMS

2.1. Calculate the resistance at 300 K of a block of Si of length 0·5 cm and cross-section $3 \times 10^{-7} \, \text{m}^2$ doped with $4 \times 10^{23} \, \text{m}^{-3}$ P atoms. What current flows when 1 V is applied across its length, and what power density is dissipated in its volume?

2.2. Plot a graph of the conductivity at 300 K of Si as the concentration of As is gradually increased from 0 to $2·0 \times 10^{22} \, \text{m}^{-3}$ in a sample originally doped with $1·0 \times 10^{22} \, \text{m}^{-3}$ atoms of B.

2.3. A photoconductor has an effective width of 10 cm and a length of 1 mm. Light of wavelength $6·274 \times 10^{-7} \, \text{m}$ falls onto it at a power density of $1·0 \, \text{W} \, \text{m}^{-2}$, generating electron-hole pairs of 100 ns lifetime. The hole mobility is $0·001 \, \text{m}^2 \, \text{V}^{-1} \, \text{s}^{-1}$ and the electron mobility is $1·0 \, \text{m}^2 \, \text{V}^{-1} \, \text{s}^{-1}$. If each photon produces an electron-hole pair, what conductance is produced by the light?

2.4. Find the carrier density for which ω_p for n-Ge equals the collision frequency as calculated from mobility and effective mass. Also, find the magnetic field for which the electron cyclotron frequency equals the collision frequency.

2.5. Find what relation there is between the formulae for the plasma frequency, the Debye length, and the root-mean-square thermal velocity of the carriers in a plasma.

2.6. A silicon surface has 10^{17} atoms/m^2 of phosphorus deposited on it. These are then diffused into the surface in such a way that they form a uniformly doped layer of thickness $2\,\mu$m. What is the surface resistivity of this layer? Ignore the conductance of the substrate. Would the answer be different if the distribution of doping atoms with depth were different? Must the distribution be known?

3. p–n junctions

Construction and direct-current characteristics

The junction between p- and n-type semiconductors has
properties which make it the basis of many electronic devices.
Carriers must be able to pass from one side of the junction to the
other without suffering recombination, so the semiconductor
must not contain more than a small number of imperfections. In
practice this means either that the device has been made from a
slice cut from a large single crystal, parts of which have been
transformed by diffusing doping atoms from the surface, or that
new material has been grown epitaxially to extend a crystal
substrate and allow the including of a p–n junction. A large p–n
junction might cover 10^{-5} m^2 (a square of side 3 mm), while in

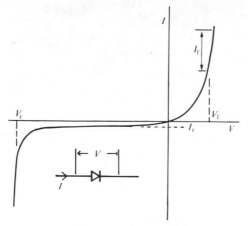

Fig. 3.1. The voltage–current characteristics for a p–n junction diode.

59

integrated circuits a device might cover no more than $10^{-10}\,\mathrm{m}^2$ area.

A p–n junction diode can rectify alternating currents, i.e. it can pass currents readily in one direction and not in the other. In Fig. 3.1, V_f, the voltage in the forward direction for large currents to start to flow is about $0\cdot2$ V for Ge and about $0\cdot6$ V for Si. I_f depends on the area and the provision for cooling the diode, and may be $10^{-4} - 10^{+2}$ A. I_r, the reverse leakage current, can be as much as a few microamperes for Ge diodes at room temperature, but is 1000 times less for Si diodes. The reverse breakdown voltage V_r is under the designer's control, and may be as low as a few volts or as high as a kilovolt.

The p–n junction in equilibrium

It is of value to analyse a simple situation—a p–n junction in equilibrium with no applied voltage or net current. Without examining the detailed variation of density or potential across the junction, three statements can be made:

(a) The net current of electrons across the junction is zero;
(b) The net current of holes across the junction is zero;
(c) At all points $pn = n_i^2$.

The zero total current density of electrons can be thought of as two cancelling components caused by drift and diffusion as in Fig. 3.2. The two components may be written as

$$-n(x)e\mu_e\frac{\mathrm{d}V(x)}{\mathrm{d}x} + eD_e\frac{\mathrm{d}n(x)}{\mathrm{d}x} = 0, \qquad (3.1)$$

where $V(x)$ is the potential and $n(x)$ the electron density at a distance x from one side of the junction (the cancellation of two currents in equilibrium is used in several places as a starting point for theory).

From eqn (3.1)

$$\mu_e\frac{\mathrm{d}V(x)}{\mathrm{d}x} = D_e\frac{1}{n}\frac{\mathrm{d}n(x)}{\mathrm{d}x}. \qquad (3.2)$$

Integrate this from $x = x_n$, well on the n-side of the junction, to $x = x_p$ safely on the p-side.

$$\int_{x_p}^{x_n} \mu_e\frac{\mathrm{d}V(x)}{\mathrm{d}x}\,\mathrm{d}x = \int_{x_p}^{x_n} D_e\frac{1}{n}\frac{\mathrm{d}n(x)}{\mathrm{d}x}\,\mathrm{d}x.$$

The left-hand side becomes an integral in V with limits V_n and V_p, and the right-hand side an integral in n with limits n_n and n_p, in each case well away from the junction. Then

$$V_n - V_p = (D_e/\mu_e)\ln(n_n/n_p).$$

If the junction was made by doping the n-side with N_d donors and the p-side with N_a acceptors, then to a good approximation

$$n_n = N_d \quad \text{and} \quad n_p = n_i^2/N_a.$$

(a) Sketch of a p-n junction in a bar of semiconductor

(b) This is an abrupt or step junction

(c) The electron-energy diagram shows that there are doping atoms in the depletion region, but few free carriers

(d) The potential variation has the opposite sign to the electron-energy variation

(e) The peak field is at the boundary between p and n

(f) This diagram can be deduced from the slope of (e) , or the addition of (b) and (g)

(g) The depletion layer is clear

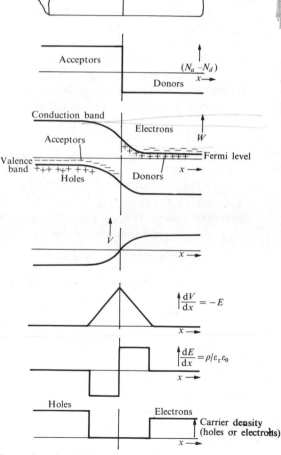

Fig. 3.2.   A p–n junction diode without bias.

Hence

$$V_n - V_p = (D_e/\mu_e) \ln(N_d N_a/n_i^2). \tag{3.3}$$

There is thus a potential difference between the two sides (referred to from now on as the *barrier potential* V_b) which curbs the tendency of the electrons to diffuse away from the place where they are densest. Unless the doping is very heavy, the barrier potential is less than the band gap of the semiconductor. Typical values of barrier potential for junctions in Ge and Si might be 0.4 V and 0.8 V.

In an equilibrium situation the Fermi level is constant right through the system. In Fig. 3.2(c) one can see that the difference of the potential energy on the two sides is equal to

$$\{(W_c - W_F) \text{ on p-side}\} - \{(W_c - W_F) \text{ on n-side}\}.$$

Equation (1.13) gives the form for this.

$$V_b = V_p - V_n = (\kappa T/e) \ln(A_c/n_n) - (\kappa T/e) \ln(A_c/n_p),$$

but, as before $n_n = N_d$ and $n_p = n_i^2/N_a$, so that

$$-V_b = (\kappa T/e) \ln(N_d N_a/n_i^2). \tag{3.4}$$

If this is compared with eqn (3.3), we can see that

$$\kappa T/e = D_e/\mu_e. \tag{3.5}$$

Equation (3.5) is known as the *Einstein relation*. The whole derivation can be repeated for holes; the barricr potential comes out the same, and there is a corresponding version of eqn (3.5)

$$\kappa T/e = D_h/\mu_h.$$

Thus if either mobility or the diffusion constant is known, the other quantity can be calculated.

Depletion layers in p–n junctions

A region close to the metallurgical junction tends to be depleted of holes and electrons, and is known as a *depletion layer* (Fig. 3.2(g)).

Figure 3.3 shows p–n junctions with forward and reverse bias. Notice that the externally applied bias shows up as a difference between the Fermi levels on the two sides—this is a fundamental

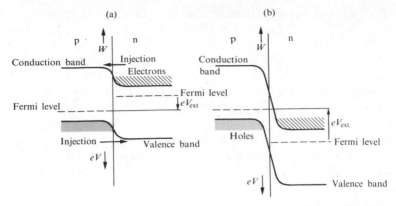

Fig. 3.3.   Energy bands for a p–n junction diode: (a) forward bias; (b) reverse bias.

point in reading or constructing diagrams. The reverse bias adds to the barrier potential and results in a wider depletion layer. The forward bias subtracts from the barrier potential—in general the forward bias never reverses the usual situation to make the p-side more positive (check with the forward bias voltages and barrier potentials quoted earlier in the chapter.)

The electron-energy diagram conveys a great deal of information about a semiconductor device, and it is well worthwhile learning how to construct these diagrams. It is not a good idea to try to memorize the diagram for every device—the simple stages of construction are what must be mastered.

(a) Start by putting the Fermi level on the paper for one of the layers of semiconductor—any one will do.

(b) Build the band round this Fermi level. The conduction band is close to the Fermi level for n-type material, but the valence band is close in p-type material.

(c) Draw the other Fermi levels at the right height on the diagram, allowing for applied voltages. The more positive of two layers is nearer the bottom of the page.

(d) Complete these bands, keeping the gap between conduction and valence bands constant.

(e) Join up the conduction band from each layer to the next, using S-shaped double curves, and do the same for the valence band.

(f) Fill in details such as free carriers, doping ions, and

applied voltages. Remember that doping ions are present in depletion layers, but that large numbers of free carriers are not.

So far we have not investigated the detailed structure of the depletion layer. This we now undertake, in a simple case, which is however a reasonable representation of some real devices. We make two assumptions. The first is that the density of doping atoms changes sharply from one value N_a on the p-side to another steady value N_d on the n-side (Fig. 3.2(b)). Such a junction is known as a *step junction* or an *abrupt junction*. In practice, as long as the change-over occurs in a distance that is much less than the full width of the depletion layer, our assumption will fit.

The second assumption is that the distance over which the electron or hole density falls to a small value is short enough to be ignored. This is equivalent to saying that the depletion layer must be much thicker than the Debye length, or that the total potential difference V_j across the junction is much greater than $\kappa T/e$.

In the depleted parts of the p-region, the only charges are fixed acceptor ions, which are negative. Thus

$$\frac{dE}{dx} = \frac{\rho}{\varepsilon_r \varepsilon_0} = \frac{-eN_a}{\varepsilon_r \varepsilon_0},$$

where ε_0 is the permittivity of free space and ε_r is the relative permittivity of the semiconductor. Hence

$$E = -(eN_a x / \varepsilon_r \varepsilon_0) + C.$$

When $E = 0$, $x = x_p$ at the boundary of the depletion layer in the p-region, so that

$$E = eN_a(x_p - x)/\varepsilon_r \varepsilon_0.$$

Because of our choice of the direction of the x axis in Fig. 3.2, x_p is a negative number, so E reaches its most negative value E_{max} when $x = 0$,

$$E_{max} = eN_a x_p / \varepsilon_r \varepsilon_0. \qquad (3.6)$$

In the depleted part of the n-region, the only charges are the donor ions, which are positive,

$$\frac{dE}{dx} = \frac{eN_d}{\varepsilon_r \varepsilon_0},$$

$$E = eN_d(x - x_n)/\varepsilon_r \varepsilon_0,$$

where x_n is the boundary of the depletion layer in the n-region. From the analysis of the n-layer,

$$E_{max} = -eN_d x_n/\varepsilon_r \varepsilon_0. \tag{3.7}$$

The total potential difference V_j across the junction is the sum of the voltages across the two parts,

$$V_j = \frac{e}{2\varepsilon_r \varepsilon_0} (N_d x_n^2 + N_a x_p^2). \tag{3.8}$$

The two equations for E_{max} (3.6) and (3.7) must give the same answer, so $N_a|x_p| = N_d|x_n|$. Hence there will be a thicker depletion layer where the doping is lighter, thinner where there is heavy doping. Remembering this we can look again at eqn (3.8) and see that most of the voltage appears across the lightly doped side. In an extreme case we can ignore the thickness and voltage on the heavily doped side. Such a junction may be indicated by a plus sign (p^+–n), and the total thickness of the depletion layer is proportional to $V_j^{\frac{1}{2}}$.

Another situation which is straightforward to analyse is a linear graded junction where the net doping varies linearly near the junction (Fig. 3.4). The density of doping atoms is then

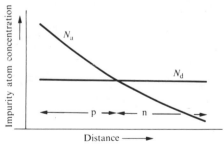

Fig. 3.4. The combination of a uniform donor concentration and a variable acceptor concentration can produce a linear graded junction.

proportional to the distance x from the metallurgical junction, E is proportional to x^2, and V to x^3. This kind of junction is often a good approximation to the junctions in planar silicon integrated circuits where doping atoms have diffused in from a surface. Instead of quoting N_a and N_d the junction is described by the gradient of the net doping density.

Depletion-layer capacitance

A p–n junction can act as a capacitor. The useful property is the small-signal capacitance $C(V_j)$, which varies with the bias voltage,

$$C(V_j) = \frac{dQ(V_j)}{dV_j} = \frac{dQ(V_j)}{dX} \frac{dX}{dV_j},$$ (3.9)

where Q is the charge on each side of the capacitor, X is the total thickness of the depletion layer, and V_j is the total difference in potential between the p- and n-sides. If we take a p–n$^+$ diode, then the major part of V_j and of X is in the p-region. Then $X \simeq x_p$ and $Q = N_a e x_p A$ (A is the area of the junction).

Hence

$$\frac{dQ}{dX} = \frac{dQ}{dx_p} = e N_a A.$$

From eqn (3.8), when $x_n \ll x_p$,

$$\frac{dX}{dV_j} = \left(\frac{dV_j}{dX}\right)^{-1} = \left(\frac{2 e N_a x_p}{2 \varepsilon_r \varepsilon_0}\right)^{-1}.$$ (3.10)

We can eliminate x_p from (3.10) by using (3.8) again and then write out the expressions for the two terms in eqn (3.9), giving

$$C(V_j) = A(e N_a \varepsilon_r \varepsilon_0 / 2)^{\frac{1}{2}} V_j^{-\frac{1}{2}}.$$

Thus $C(V_j)$ decreases as the bias becomes more negative. If $C(V_j)$ is expressed in terms of the depletion-layer thickness (check this using eqn (3.8) once more), then

$$C = A \varepsilon_r \varepsilon_0 / X,$$ (3.11)

so the capacity is identical with that of an ordinary capacitor of the same size, shape, and permittivity as the depletion layer.

This is true for all p–n junctions, no matter what the variation of doping with position. To see this, consider a p–n junction with a voltage V across a depletion layer of total thickness X. If the depletion layer is expanded by including a charge layer dQ, then the change in field dE on passing through dQ is dQ/ε. This change in field contributes a total change in voltage over the full width of the depletion layer of $X\,dE$, so that

$$dV = X\,dE = X\,dQ/\varepsilon$$

and hence

$$dQ/dV = \varepsilon/X = C(V_j).$$

The change in field is eventually cancelled by the corresponding charge layer $-dQ$ on the other side of the junction.

In our analysis of transistors we shall see how the depletion-layer capacitance is a factor which limits performance. In most diodes the capacitance is unwanted, but in varactor diodes or varicaps, the capacitance is exploited.

Some circuits (including parametric amplifiers) require a capacitor whose value can be controlled by an electrical signal. A reverse-biased p–n junction will do this job admirably—when the capacity is to vary, the bias can be altered. The maximum frequency f at which a varactor diode can be used is set by the time constant $\tau = C(V_j)R$, where R is the resistance of the semiconductor on either side of the p–n junction,

$$f \ll \frac{1}{2\pi\tau}.$$

If a linear graded junction is used instead of a step junction, the relation between capacity and voltage takes the form

$$C(V_j) \propto V_j^{-\frac{1}{3}}, \tag{3.12}$$

and other forms (which may be convenient for use in particular circuits) can be obtained by suitably arranging the variation of doping density.

Minority-carrier injection

When a p–n junction is forward-biased (p made more positive), there is a lowering of the barrier which was preventing the majority carriers from moving across the depletion layer. Consequently there will be a flow of majority carriers across the junction, becoming minority carriers and contributing to a measurable current. We shall first examine conditions at the depletion-layer boundary, and then the way the carriers move away from the junction.

Depletion layers are thin, and carriers can move frequently backwards and forwards across them. As a result there is a local equilibrium between the carrier densities on the two sides of a depletion layer for each kind of carrier separately. If we use a subscript p or n to indicate the location of carriers and a subscript 0 to indicate equilibrium (e.g. p_{n0} means the density of holes in the n-side in equilibrium), then the relation between densities on the two sides of the junction is

$$p_n/p_p = \exp(-eV_j/\kappa T) = n_p/n_n, \tag{3.13}$$

which is another application of the Boltzmann relation.

Because neutrality is maintained outside the depletion layers, the majority and minority-carrier densities increase (or decrease) by equal amounts. As a result, while the minority-carrier density can readily change by a large factor, the majority-carrier density changes only by a small factor, and in many circumstances can be assumed to be the equilibrium value. When this is so

$$p_n/p_{p0} = \exp(-eV_j/\kappa T) = n_p/n_{n0}. \tag{3.14}$$

If we use the semiconductor equation to provide an expression for the minority-carrier density, and substitute this in eqn (3.4), we can obtain

$$p_{n0}/p_{p0} = \exp(-EV_b/\kappa T) = n_{p0}/n_{n0}, \tag{3.15}$$

The applied voltage $V_{ext} = V_j - V_b$, so that, combining eqns (3.14) and (3.15)

$$p_n/p_{n0} = \exp(-eV_{ext}/\kappa T) = n_p/n_{p0}. \tag{3.16}$$

Put into words: the minority-carrier density on one side of a p–n junction is physically related to the majority-carrier density on the other side by a term containing the total potential difference between the two sides. But if we like we may relate the perturbed carrier density to the equilibrium carrier density by an exponential term containing the externally applied voltage. This

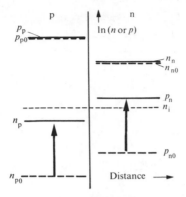

Fig. 3.5. Carrier densities across a p–n junction with forward bias. On a ln(density) scale, n_i comes midway between the majority and minority carrier densities, and both minority-carrier densities are raised by the distance $\ln\{\exp(eV_{ext}/\kappa T)\}$ when a voltage V_{ext} is applied.

is often very convenient, as the applied voltage is easy to measure, while V_j is not.

Figure 3.5 shows the relations pictorially. When the density is plotted on a logarithmic scale, then both the minority-carrier densities will be moved the same distance on the graph away from the equilibrium condition. An important point is that more minority carriers come from the more heavily doped side—this is exploited in the base-emitter junction of a bipolar transistor discussed on pp. 93–116.

Voltage–current relation for a p–n junction diode

By combining the results of the previous section with those of (1.17), we can describe the flow of carriers across and away from the depletion layer in a p–n junction diode. We consider a diode which is much thicker than the diffusion lengths L_e or L_h (Fig. 3.6). The continuity equation (1.26) for the excess electron density n' is

$$D_e(\mathrm{d}^2 n'/\mathrm{d}x^2) - n'/\tau = 0. \tag{3.17}$$

This is simpler than eqn (1.26), because the electric field is low enough to ignore if the currents are not too large, the equilibrium generation and recombination terms have cancelled

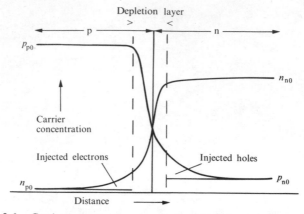

Fig. 3.6. Carrier concentrations in a forward-biased p–n junction diode.

out, and the total rate of change of density is zero in a steady situation. In the p-side we can write the excess density as a function of the distance x from the edge of the depletion layer, $n_p'(x)$.

From eqn (3.16), the excess density at $x = 0$ is

$$n_p'(0) = n_{p0}\{\exp(eV_{ext}/\kappa T) - 1\}.$$

The excess density falls to zero as $x/L_e \gg 1$, so the solution to eqn (3.17) is

$$n_p'(x) = n_p'(0) \exp(-x/L_e). \qquad (3.18)$$

From eqn (3.16), the excess density at $x = 0$ is

$$n_p'(0) = n_{p0}\{\exp(eV_{ext}/\kappa T) - 1\}.$$

The current density of the diffusing electrons is

$$j_e(x) = eD_e \, dn_p'(x)/dx,$$

and at the edge of the depletion layer ($x = 0$) it is

$$j_e(0) = (eD_e n_{p0}/L_e)\{\exp(eV_{ext}/\kappa T) - 1\}.$$

A similar analysis for holes injected into and flowing in the n-side gives a similar expression for the current density carried across the depletion layer by holes. The total current density j is the sum of the two contributions

$$j = j_e(0) + j_h(0)$$
$$= e\{(D_e n_{p0}/L_e) + (D_h p_{n0}/L_h)\} \{\exp(eV_{ext}/\kappa T) - 1\}. \quad (3.19)$$

The first part of the equation is a constant for a given diode, and can be written as j_r, so the whole equation then becomes

$$j = j_r\{\exp(eV_{ext}/\kappa T) - 1\}.$$

For a diode of area A, the total current I will be jA, and if we define $I_r = Aj_r$, the d.c. voltage–current relation for a p–n junction diode is

$$I = I_r\{\exp(eV_{ext}/\kappa T) - 1\} \quad (3.20)$$

(check that the equation behaves in a suitable way to fit Fig. 3.1). The expression for I_r can be written out in terms of the doping density rather than the minority-carrier density.

$$I_r = Aen_i^2\{(D_e/L_e N_a) + (D_h/L_h N_d)\} \quad (3.21)$$

For many purposes a p–n junction is required to have I_r small. Inspection of eqn (3.21) shows what is needed to achieve this—in particular n_i can be made small by choosing a material with a high band gap, and by avoiding high temperatures, and N_a and N_d should both be large.

The theory leading to eqn (3.20) is an accurate description of some p–n junction diodes for moderate forward currents and reverse voltages. For other diodes other effects are important in addition to those included in our analysis. For instance, the resistance of thick semiconductor layers may increase the total voltage of the diode for a given current; the depletion layer may be wide enough for a noticeable number of electron-hole pairs to be generated within it; the minority-carrier densities may rise to approach the majority-carrier densities. Theories including some or all of these more advanced effects can be found elsewhere (see Sze 1981; Yang 1978), but the Boltzmann equation, minority-carrier injection, and diffusion will always form a basis for understanding the physical processes occurring in p–n junction diodes.

The small-signal differential resistance r of a diode can be obtained by differentiating eqn (3.20). It has a simple form when

$\exp(eV_{\text{ext}}/\kappa T) \gg 1$:

$$r = \frac{dV}{dI} = \left(\frac{dI}{dV}\right)^{-1} = \frac{\kappa T}{eI}. \tag{3.22}$$

At 300 K, $\kappa T/e = 1/40$ V, so that if I is in amperes the value of r in ohms is $1/(40I)$.

This is a useful formula for describing the input impedance of a bipolar transistor in terms of its base current.

Diffusion capacitance

When a p–n junction diode is passing current, there are more electrons and holes in it than when the current is zero. These charges have to be supplied, and as a result the diode acts as a capacitor with a capacitance C_d, additional to $C(V_j)$.

Define C_d as dQ/dV. For this problem it is convenient to think of the relation of the extra charge Q to the diode current I, so we write

$$C_d = dQ/dV = (dQ/dI)(dI/dV).$$

The value of Q is found by integrating the excess charge density described by eqn (3.18). For the electrons in the p-side,

$$Q_e = \int_0^\infty eAn_p(x)\,dx = eAn_{p0}\{\exp(eV/\kappa T) - 1\}\int_0^\infty \exp(-x/L_h)\,dx.$$

Hence

$$Q_e = eAn_{p0}L_e\{\exp(eV/\kappa T) - 1\}.$$

There will be a similar expression for the excess holes in the n-side, so that the total excess charge is

$$Q = eA\{\exp(eV/\kappa T) - 1\}(n_{p0}L_e + p_{n0}L_h)$$

or, for compactness,

$$Q = Q_0\{\exp(eV/\kappa T) - 1\}.$$

Thus, from eqn (3.20), we can write

$$Q = Q_0 \cdot I/I_r, \qquad dQ/dI = Q_0/I_r.$$

We already have a form for dI/dV from eqn (3.22) so that

$$C_d = (eI/\kappa T)(Q_0/I_r). \tag{3.23}$$

The diffusion capacitance increases as the forward current increases, becoming larger than the depletion layer capacitance for all reasonable forward biases. It is always associated with the diode differential resistance r, so the forward-biased diode is inevitably lossy, and can only be used to store charge for times $\ll \tau$. A full analysis shows that the impedance of a forward-biased diode varies with frequency ω as

$$Z(\omega) = 1/(1 + j\omega^2 \tau^2)^{\frac{1}{2}}.$$

Thus, the speed of the diode's response depends on the recombination time of its carriers.

Breakdown in p–n junctions

Breakdown in p–n junctions is the rapid increase of current as the reverse bias voltage is increased beyond some safe limit. Breakdown may result in the destruction of a device but this is not inevitable. Damage is caused by local melting, and if the current is held to a small value by an external resistance, or is spread over a large area by careful design and manufacture, then the temperatures reached are not high enough for silicon or SiO_2 to melt, nor for impurities to diffuse out of their intended positions.

Figure 3.1 shows a current–voltage plot which defines the breakdown voltage V_r. V_r depends on the doping density of a diode, in particular on the doping of the more lightly doped side of the junction. Figure 3.7 shows a plot of breakdown voltage for diodes of different doping. It indicates the general value of the quantities, but does not account for the grading of the junction, the doping of the more heavily doped side, nor the distinction between a planar and a spherical junction.

There are two distinct breakdown mechanisms—avalanche breakdown at high voltages and Zener or tunnel breakdown at low voltages for heavily doped diodes. As the doping density rises above 10^{25} the Zener breakdown voltage falls to zero, and the diode shows no rectifying property.

Zener breakdown

Zener breakdown occurs where electrons in the valence band find themselves only a short distance away in real space from

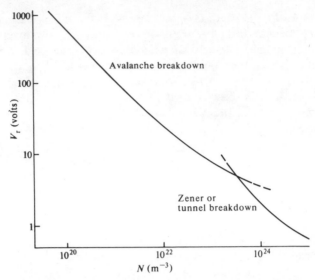

Fig. 3.7. Breakdown voltage V_r versus doping density N for a Si junction diode. The breakdown voltages for diodes of other materials are roughly in proportion to the ratio of their band gaps.

empty states in a conduction band. Although they have insufficient energy to overcome the potential in the forbidden band, they can tunnel through the barrier if it is short, and emerge into existence on the far side. The height of the energy barrier is the band gap, and the tunnelling distance may be calculated from the formula for the shape of a depletion curve. If a doping density of $10^{24} \, \text{m}^{-2}$ is taken from Fig. 3.7, the width W_T of the forbidden region may be found by calculating the position of the edge of the conduction band at two potentials which differ by W_G/e as in Fig. 3.8.

$$W_T = x_1 - x_2 = \left(\frac{\varepsilon V_2}{eN_a}\right)^{\frac{1}{2}} - \left(\frac{\varepsilon V_2}{eN_a}\right)^{\frac{1}{2}}.$$

The tunnelling distance is found to be about 20 Å or about 7 atoms. For greater distances the chance of tunnelling becomes too small to be of practical effect.

A characteristic feature of Zener breakdown is that it occurs more easily as temperature rises. This is the opposite of the effect of temperature on avalanche breakdown.

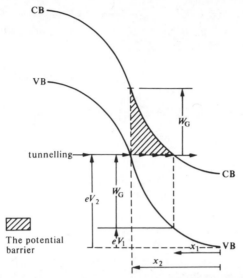

Fig. 3.8. Breakdown by tunnelling. The potential barrier is shown shaded, and has a maximum height of W_G/e. The width of the barrier for an electron tunnelling from valence band to conduction band can be found from the thickness of the depletion layer at energies eV_2 and eV_1 above the lowest energy in the band, where $eV_1 - eV_2 = W_G$.

Avalanche breakdown

In avalanche breakdown, carriers gain energy from the electric field in the junction and make ionizing collisions which release new electron–hole pairs. If the process occurs to a moderate extent and is under control the term 'avalanche multiplication' is used. If the multiplication is large and the resulting current is not controlled by the initial current flowing into the junction, then the term 'avalanche breakdown' would be more suitable.

For a carrier to make a collision and make a new electron–hole pair, it needs kinetic energy of at least twice the band gap. This is demanded by the need to conserve both momentum and energy, and can be shown by a calculation assuming that electrons behave like conventional billiard balls, but losing energy W_G on impact. The centre-of-mass frame provides a simple derivation.

The breakdown voltage is found to be higher for diodes with

lower doping. To a first approximation, breakdown occurs when the field in the depletion layer rises above a critical level E_{br}.

For a symmetrical abrupt diode

$$\frac{V_j}{2} = \frac{eNX^2}{2\varepsilon} \quad \text{and} \quad E_{max} = eNX,$$

where X is the width of one side of the depletion layer so that

$$V_r = \frac{1}{N_a} \frac{E_{br}^2}{e\varepsilon}.$$

Thus the notion that breakdown occurs when the peak field rises above E_{br} leads to the prediction of V_r varying inversely with doping density. The value found for E_{br} is roughly $3 \times 10^7 \, \text{V m}^{-1}$ for Si.

By contrast with Zener breakdown, the breakdown voltage for avalanche breakdown increases with increasing temperature. When the two mechanisms are contributing equally at about 5 to 7 V, the breakdown voltage is nearly independent of temperature and particularly stable reference voltage devices (Zener diodes) can be made.

Devices for optical fibre communications

Optical fibre communication systems transmit signals tens of kilometres over silica fibres, which have their maximum transparency at $1{\cdot}3 \, \mu\text{m}$ and $1{\cdot}55 \, \mu\text{m}$, corresponding to $1{\cdot}0$ and $0{\cdot}8$ eV. There is thus a need for transmitters and receivers at these wavelengths. The transmitters that have been found most effective are semiconductor LEDs and lasers, which are discussed later in this chapter. Two receivers are p–i–n diodes, and avalanche photo-diodes.

The p–i–n diode is a modified p–n junction whose properties have been optimized to act as a sensitive, fast light detector. The central layer (Fig. 3.9) is n-type silicon, so lightly doped that it is often thought of as intrinsic. Because it is lightly doped, when it forms part of a depletion layer any space charge is low, and electric field changes with position are small. The result is that a moderate reverse bias can produce a thick depletion layer (perhaps 50 μm), as has been explained after eqn (3.8).

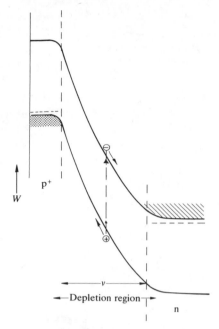

Fig. 3.9. The large reverse bias on a p–i–n (or p–v–n) photodiode produces depletion through the whole of the v layer. Photons absorbed there produce carriers that are swept away to the terminals.

Such a depletion layer is a suitable thickness for absorbing photons emerging from an optical fibre. The electric field ensures that the electron–hole pairs separate and contribute to an external current, and also that the transit time through the depletion layer is short. If 10 V is applied across 50 μm, the average field is $2 \times 10^5 \, \mathrm{V \, m^{-1}}$, the mean drift velocity is $2 \times 10^4 \, \mathrm{m \, s^{-1}}$, and the transit time is about 2·5 ns. The area of a p–i–n diode needs to be just large enough to accept the light from an optical fibre, in contrast to the large area of solar cells.

Solar cells

Solar cells are p–n junction diodes of large area whose properties are optimized for the absorption of sunlight, and the collection of the resulting photo-electrons and holes. In Fig. 3.10(a), sunlight passes through the SiO_2 coating, which not only protects the silicon from contamination, but has a suitable

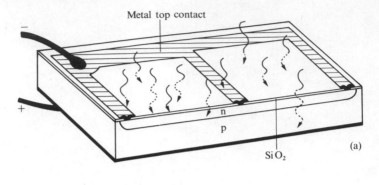

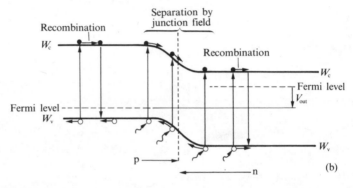

Fig. 3.10.   A solar cell: (a) note the top contact design, and the penetration of photons to various depths; (b) the energy-band diagram for a solar cell. Only those electron-hole pairs which separate across the depletion layer contribute to the output current.

refractive index for reducing the amount of light reflected from the surface. Light is absorbed through the volume of the silicon, but only some of the photons or holes cross the depletion layer and contribute to the external current The photo-current is independent of the bias applied to the cell, and the process of minority carrier injection and diffusion still occur, so the equivalent circuit for a solar cell is a p–n junction diode having a dark reverse saturation current I_r, with a light-dependent current source I_L in parallel with it. The resistance R_c of contacts can be important so this is also shown on the equivalent curcuit (Fig. 3.11(a)).

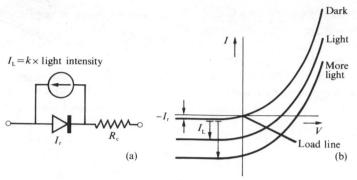

Fig. 3.11.   Solar cell: (a) equivalent circuit; (b) the voltage–current characteristics show a downward shift as the light becomes stronger.

The current I through a solar cell is given by

$$I = I_r\{\exp(eV_s/kT) - 1\} - I_L,$$

while the voltage V_s measured across the semiconductor is related to the output voltage V_{out} by

$$V_{out} = V_s - IR_c.$$

The current–voltage characteristics are shown in Fig. 3.11(b) for a cell where R_c is small. When the cell is in use, it is connected to a load that absorbs power, and will operate in the second quadrant of the graph. The short-circuit current I_{sc} and the open-circuit voltage V_{oc} are

$$I_{sc} = I_L \text{ (if } R_c \text{ is small)}$$

$$V_{oc} = \frac{kT}{e} \ln\left\{\frac{I_L + I_r}{I_r}\right\}.$$

The design of the contact to the top surface (Fig. 3.10(a)) has to be a compromise between the need for low resistance, and the need to avoid blocking the path of the sunlight. Another compromise is made in selecting a semiconductor with the best band gap. Only those photons with energy greater than the band gap of the semiconductor can produce electron–hole pairs, so a low band-gap material has an advantage as more of the incident photons generate photo-current. On the other hand, as the band gap increases, so does the output voltage, so that high efficiency of conversion of solar power to electrical power is predicted to come from the use of a semiconductor of band gap 1·2 to 1·8 eV. Both GaAs and Si are suitable, and most solar cells are made of silicon.

Further conditions on the properties of the semiconductor can be understood by examining Fig. 3.10(a) and (b). The light must be absorbed before it passes right through the semiconductor, so the average absorption length l(abs) must be about the thickness of the p-layer, i.e. 10–20 μm. Next, the photo-electrons and holes must be transported across the depletion layer before they recombine. Collection occurs from the thickness X of the depletion layer plus about one diffusion length on either side, so that another design criterion is

$$L_e + L_h + X \simeq l(\text{abs}).$$

The requirements on L_e and L_h for a solar cell are less severe than when making high-gain transistors, and the possibility of using less pure silicon arises. Polycrystalline or amorphous silicon both require less energy to produce, and could be much cheaper than the single crystal slices used for transistors and integrated circuits. In many cases it is true that sunlight is free, so the important figure of merit of a solar cell is the cost/watt, not the efficiency, and the widespread use of solar cells may not come about until an alternative is found for expensive high quality silicon.

High-energy particle detectors

Semiconductor diodes are used to detect high-energy particles such as α-particles or fast protons or deuterons. The fast particles are slowed to rest in a thick diode, often of silicon, and their energy is partly converted to heat, and partly used to produce electron–hole pairs. The electron–hole pairs are separated by the electric field in a depletion region in the same way as in a photodiode with reverse bias across it.

A pulse of current flows each time an ionizing particle is stopped by the detector. In silicon an α-particle makes one electron–hole pair for each 3·6 eV it loses. Consequently a 5-MeV α-particle (a typical energy) produces $5 \times 10^6/3\cdot6 = 1\cdot5 \times 10^6$ electron–hole pairs, which could be collected in time t

governed by the width of the depletion layer—typically 10 ns. The resulting current pulse I is given by $I = Ne/t$ so that

$$I = 1 \cdot 5 \times 10^6 \times 1 \cdot 6 \times 10^{-19}/10^{-8} = 2 \times 10^{-5} A.$$

This is easy to amplify and measure.

A more energetic α-particle produces more electron–hole pairs, and a large current pulse, so that the energy of the α-particle can be deduced from the size of the pulse.

The next question is—do all the α-particles of the same energy produce exactly the same number of electron–hole pairs? The answer is that there is some variation, corresponding in the case of a 5-MeV α-particle to about ± 7 keV in the measured α-particle energy.

A simple argument predicting this effect assumes that an α-particle of energy W loses its energy in W/P packets each of energy P, where P is half the average energy per electron–hole pair. Half the packets produce an electron–hole pair; the other half fail to do so, and their energy heats the crystal lattice. The mean number $\bar{N}$ of electron–hole pairs is $W/2P$, as it should be. Because the possible values of N form a binomial distribution, the standard deviation of N is $(N \times \frac{1}{2} \times (1 - \frac{1}{2}))^{\frac{1}{2}}$.

For the 5-MeV α-particle in silicon, N is about $2 \cdot 4 \times 10^6$, so the standard deviation of the calculated energy is about $800 \times 3 \cdot 6$ eV or about 3 keV. This is rather lower than the experimental figure. The difference can be explained if the energy is lost in larger packets, which produce several electrons together.

The range of a 5-MeV α-particle in silicon is about 30 μm. If a depletion layer of this width is to be formed in silicon with a bias voltage of 100 V, then a doping density of about 10^{20} donors or acceptors per m^3 is needed. This is a low value, and even lower values are used in practice, in contrast to the much higher values in transistors.

Light-emitting diodes

LEDs and infra-red emitting diodes are p–n junction diodes whose construction and material are selected to produce light when current is passed through the diode in the forward direction. Electrically, the theory of pp. 69–72 applies without modification; for instance voltage–current relations will be of the form $I = I_r\{\exp(eV_{ext}/\kappa T) - 1\}$, and in use very little current will flow until V_{ext} is more than some minimum value. For diodes emitting visible light, a forward bias of 1·5 V–2·5 V is usually required. The intensity of the light output is found to be proportional to the current.

The basic process occurring is the emission of a photon when an injected electron (or hole) recombines, so in considering the physics of LEDs, attention is focused on to the possible recombination processes in semiconductors and the way in which the recombination energy is dissipated. Most recombinations occur within two diffusion lengths of the junction, so the light will emerge from this region. One possible transition is directly from the conduction band to the valence band (Fig. 3.12(a)) when the energy at the peak of the spectrum will be about equal to the band-gap energy $+\kappa T$. Other transitions may occur by stages, the energy of any photon emitted being less than the band-gap energy (Fig. 3.12(b), (c), (d)). We can thus recognize two ways of controlling the energy of emitted photons. We can select a semiconductor with the right band gap, or we can add impurities to provide energy levels in the forbidden band. In practice both methods are employed.

A further problem is to arrange that a high enough fraction of the transitions go by a route where a photon is emitted, and that non-radiative transitions are few. The distinction between a direct-gap and an indirect gap semiconductor is important here.

In a direct-gap semiconductor (e.g. GaAs, InSb) the minimum of the conduction band and the maximum of the valence band are both at the same value of k, so that transitions such as α in

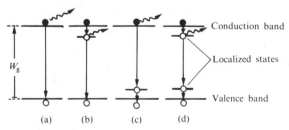

Fig. 3.12.   Transition processes for electrons. The photon is shown leaving the initial state of the radiative transition.

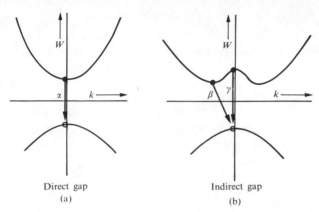

Direct gap
(a)

Indirect gap
(b)

Fig. 3.13.   Electron transitions shown on an electron energy versus wave-number diagram for semiconductors having a direct gap and an indirect gap.

Fig. 3.13(a) require no change of momentum, and all the released energy can be carried away be the low momentum photon. In an indirect-gap semiconductor (e.g. Si, Ge, GaP), the minimum of the conduction band is not at $k = 0$, so that a transition such as β in Fig. 3.13(b) must involve some other particle to conserve momentum. The other body is usually a phonon, which also carries away the energy, so the transition is non-radiative. Transitions like γ are likely to be rare as there will be few electrons occupying the high energy $k = 0$ region of the band.

A photon is only useful when it has emerged from the diode. Absorption in the semiconductor, and total internal reflection at the surface reduce the effective output by a factor of up to one hundred in some devices.

Some examples from the many types of light-emitting diodes are as follows:

(a) GaAs infra-red-emitting diodes. These emit at 1·4 eV in the near infra-red by a band-to-band transition across the 1·4 eV direct band gap of GaAs. The internal quantum efficiency (fraction of injected electrons recombining with emission of a photon) may reach 15 per cent, but the external efficiency (infra-red power out/electrical power in) is only about 1 per cent because the photons are readily reabsorbed.

(b) GaP doped with Zn and O. These emit red light of 1·7 eV energy. The band gap of GaP is 2·25 eV, and is indirect. The Zn and O, when on adjacent lattice sites, form an unusual neutral impurity energy level. The external efficiency is about 2 per cent because the chance of reabsorption is low.

(c) GaP doped with N. These emit a broad band of light whose appearance to the human eye changes from red to green as the amount of N is decreased. The N may form a band of states in the forbidden energy gap which broadens as the N atoms come closer together. Efficiency is low, but as the eye is sensitive to yellow and green, the brightness is acceptable.

(d) GaAs$_x$P$_{1-x}$ doped with moderate amounts of N. As long as x is less than 0·5 the band structure is that of GaAs with a direct gap (Fig. 5.18). The band gap increases as the fraction of P increases, and at the same time the colour of the light changes from red through to green.

Lasers

In this section we shall consider the physics of lasers, and then the semiconductor junction–diode lasers that have been developed from LEDs.

We start with the idea that the transition of an electron between two energy levels S_1 and S_2 can occur in three ways, ignoring the possibility of interactions with other particles or with quasi-particles such as phonons.

(a) Upwards, by absorption of a photon, at a rate given by

$$dN_2/dt = B_1 U(v)N_1,$$

where B_1 is a constant, N_1 and N_2 are the density of

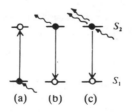

(a)   (b)   (c)

occupied S_1 and S_2 levels, and $U(v)$ is the density of radiation of frequency v.

(b) Downwards, by spontaneous emission of a photon, at a rate

$$dN_2/dt = -AN_2.$$

(c) Downwards, by stimulated emission of a photon. This happens when an excited state is kicked into emission by the field of a passing photon. The transition rate is

$$dN_2/dt = -B_2 U(v)N_2.$$

The emitted photon is a replica of the photon which stimulated it.

The total rate of change of N_2 is

$$dN_2/dt = -AN_2 + B_1 U(v)N_1 - B_2 U(v)N_2.$$

In a steady state $dN_2/dt = 0$, and hence

$$U(v) = \frac{A}{B_1(N_1/N_2) - B_2}.$$

One example of a steady state is thermodynamic equilibrium, where the Boltzmann relation $N_1/N_2 = \exp(-eV/\kappa T)$ relates N_1 and N_2 to eV, the difference in energy between S_1 and S_2. Thus in thermodynamic equilibrium

$$U(v) = \frac{A}{B_1 \exp(eV/\kappa T) - B_2}.$$

However, this must be equivalent to the Planck formula for radiation density in any equilibrium situation at a temperature T.

$$U(v)\,dv = \frac{8\pi h v^3}{c^3} \frac{1}{\exp(hv/\kappa T) - 1}\,dv.$$

If the two formulae are to match, $B_1 = B_2 (=B)$ and $A/B = 8\pi h v^3/c^3$. Because $B_1 = B_2$ the probability of a transition that depends on the presence of a photon is the same in both directions, the actual ratio being determined by N_1/N_2. The ratio A/B is proportional to v^3, and hence at low (microwave) frequencies spontaneous transitions are rare, while at high (visible) frequencies, stimulated transitions are relatively less frequent, and lasers more difficult to operate.

In a laser, two techniques are used to make the stimulated or downward transition the dominant process.

(a) Population inversion, which means arranging for a high energy state to be more fully occupied than some state of lower energy. As a result $BU(v)N_2 > BU(v)N_1$, so that more photons are emitted by stimulated emission than are absorbed. To achieve population inversion requires a major shift away from equilibrium, and a high energy flow.

(b) Photon confinement, so that high photon densities can be reached. When $U(v)$ is high enough $BU(v)N_2 > AN_2$, and stimulated emission becomes more frequent than spontaneous emission.

A semiconductor junction–diode laser achieves population inversion by strong forward bias of a very heavily doped p–n junction, and a high photon density by reflection between two opposite polished faces of the semiconductor chip. The critical transition can be any of those indicated in Fig. 3.11.

The requirement for population inversion in a semiconductor must take into account the spread of energies of the relevant states near the edges of the conduction and valence bands. As on pp. 27–9, the rate of transition between the two bands is proportional to the number of candidates ready to make a transition, and to the number of places they may fill.

The rate of transition upward is

$$f_v A_c (1 - f_c) A_c U(v) B,$$

while the transition rate downward by stimulated emission is

$$f_c A_c (1 - f_v) A_v U(v) B,$$

where A_v and A_c are the effective densities of states in the valence and conduction bands, and f_v and f_c are the fractions of the states filled by electrons. We are assuming that $U(v)$ is high enough to allow spontaneous transitions to be ignored. No exponential factor distinguishes upward and downward transition rates as the same photon density drives both. If stimulated emission is to be more frequent than absorption then

$$f_v A_v (1 - f_c) A_c < f_c A_c (1 - f_v) A_v$$

or

$$f_c > f_v.$$

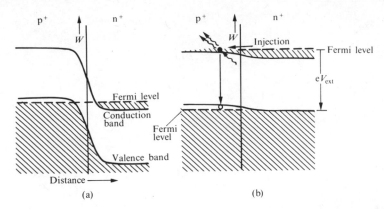

Fig. 3.14. Energy bands for a junction–diode laser: (a) without bias: (b) with enough bias to invert the electron population in the p^+ region near the junction.

Thus for the laser to operate, there have to be more electrons near the bottom of the conduction band than near the top of the valence band. This can be achieved by doping the junction so heavily that the Fermi level is above the bottom of the conduction band in the n-side, and below the top of the valance band in the p-side (Fig. 3.14(a)), and applying a forward bias about the size of the barrier potential which will be a little more than the band gap, as in Fig. 3.14(b), which represents a GaAs laser with an infra-red output. The current density of such a laser at 300 K is 6×10^8 A m^{-2}; a chip 0·1 mm square would require 6 A.

By using a three-layer structure, where only the meat in the sandwich is GaAs, the threshold current density is reduced to about 5×10^6 A m^2 (Fig. 3.15). The outer two layers are Al$_x$Ga$_{1-x}$As, which has a band gap of 1·9 eV by comparison with 1·4 eV for GaAs. The p-Al$_x$Ga$_{1-x}$As confines the injected electrons to the p-GaAs, so that an adequate density can be built up with a lower current. A second effect comes from the lower refractive index of the two Al$_x$Ga$_{1-x}$As layers, which help to confine the photons to the central active region. Figure 3.16 shows a sketch of a typical GaAs/GaAlAs narrow-stripe double hetero-junction laser diode. Points to note are:

(a) The active region is sandwiched between layers of higher band gap, to confine charge carriers, and of lower

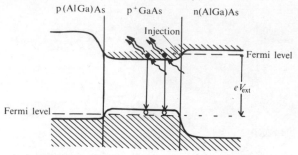

Fig. 3.15. A double-heterojunction laser with working bias. Injected electrons are trapped in the p⁺ GaAs conduction band.

refractive index, to confine photons. The values of W_g and n have been found from a weighted average of the values for GaAs and AlAs.

(b) The layers of semiconductor all have nearly the same lattice constant, so that epitaxial growth is possible.

(c) The flow of current is confined to a narrow stripe by a SiO_2 layer. Several other techniques have been devised to do this.

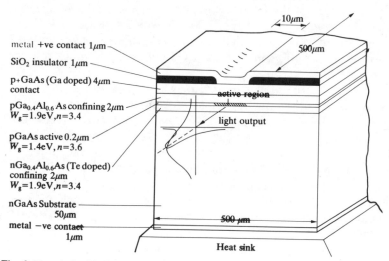

Fig. 3.16. A double-heterojunction narrow-stripe laser diode, that could be run continuously at room temperature.

(d) The upper three layers of semiconductor are p-type and the lower two are n-type, so the whole block has just one p–n junction in it.

(e) The beam that emerges is narrow horizontally and broad vertically. A cylindrical lens can produce a narrow beam. The active area of the laser is as small as the core of many optical fibres.

Figure 3.17 shows the threshold current for a laser diode. In use a laser diode is biased just below the threshold current, and then rapidly switched into the lasing condition.

Optical fibre communications need an optical source with the following characteristics.

1. An output wavelength which matches an absorption minimum in silica fibre at 1·1 or 1·5 μm. GaAlAs can give the shorter wavelength, and quaternary compounds are used for the longer wavelength.

2. Narrow output spectrum to reduce the effect of dispersion in the fibre. Semiconductor lasers can have a line width of less than 1 per cent.

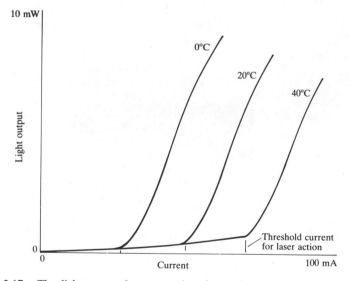

Fig. 3.17. The light output from a semiconductor laser rises rapidly once a threshold current is exceeded, and laser action occurs. The threshold current increases with ambient temperature.

3. High output power over a small area to permit a long transmission distance before the signal falls below the threshold of the receiver. Semiconductor lasers can launch 1 mW into a fibre, which is enough for a link of 20 km (Fig. 3.17).

4. Rapid switching time for a high information transmission rate. Switching speeds of less than 1 ns are common, allowing data rates above 1 Gbit s^{-1}.

5. A long life so that devices do not need replacing in inaccessible equipment. The intense beam of light tends to damage the reflecting surfaces of the laser, but mean lives of up to 10^7 h are now possible.

IMPATT diodes

It has been found that many kinds of diodes which embody a p–n junction can act as microwave oscillators when placed in a resonant circuit and reverse-biased just beyond the edge of avalanche breakdown. The name 'IMPATT diode' derives from a fuller title IMPact Avalanche Transit Time diode.

One of these diodes is sketched in Fig. 3.18(a). The total field takes into account both the barrier potentials and the applied voltage (Fig. 3.18(b)). In the central n-region, the field is somewhat above the value E_{dr} required to make the electrons

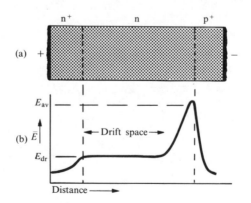

Fig. 3.18. IMPATT diode: (a) schematic structure: (b) variation of average electric field E along the diode.

drift at their maximum speed, while at the n–p$^+$ junction, the field rises above the value E_{av} required for an electron–hole avalanche to start. The device is in a resonant circuit, so there will be a repeating cycle of events.

After the peak field rises above E_{av} the avalanche builds up, reaching its maximum only when E has passed its own peak and fallen again below E_{av}. The holes produced in the avalanche rapidly reach the p$^+$-contact, taking no further part in the process, but the electrons are released into the n-region where they are not neutralized by either donors or holes. As a result the field must rise where the injected electrons are present, and there is less voltage difference to develop a high field across the n–p$^+$ junction, so that the avalanche ceases. The electrons drift at their maximum velocity across the n-region, and current continues to flow in the external circuit while they do so. Thus there are two delays between the voltage and the current waveforms, and if the frequency of the resonant circuit is suitable, the current can be between $\frac{1}{2}\pi$ and $\frac{3}{2}\pi$ out of phase with the voltage. It is therefore feeding a.c. power to the circuit rather than absorbing it as a normal resistive load does.

To make a high frequency, the drift length and avalanche delay must be short. Power output of 2 W at 7 GHz with a d.c. power supply of 100 V and an efficiency of 6·5 per cent is available commercially, while 200 μW at 361 GHz has been achieved in experiments.

Other modes of oscillation are seen in practice, some of which are of much higher efficiency and hence of interest to potential users.

PROBLEMS

3.1. An abrupt Si p–n junction is doped with $10^{21}\,\mathrm{m}^{-3}$ B atoms in the p-region and $10^{20}\,\mathrm{m}^{-3}$ P atoms in the n-region. Calculate, at 300 K:

    (a) V_b, the barrier potential;

    (b) the depletion-layer thickness in each side when -10 V is applied;

    (c) the capacitance with -10 V bias, if the area is $10^{-8}\,\mathrm{m}^2$;

    (d) the minimum voltage for avalanche to start.

(For part (d), assume the avalanche occurs when the field exceeds $3 \cdot 0 \times 10^7 \, \text{V m}^{-1}$ at some point.)

3.2. A reverse-biased p–n junction diode may be used as a voltage-controlled capacitor in a tuned circuit where the resonant frequency f is proportional to (diode capacitance)$^{-\frac{1}{2}}$. If a symmetrical Si step junction diode has a nett doping density on each side of the junction of $10^{-22} \, \text{m}^{-3}$, by what factor does f vary as the bias is changed from $-1 \, \text{V}$ to $-10 \, \text{V}$?

3.3. A Si step junction is initially unbiased. A current of 1 mA is passed through it, so as to make it steadily more negatively biased. How long does it take for the bias voltage to reach $-10 \, \text{V}$? Take the doping density on each side of the junction to be $10^{21} \, \text{m}^{-3}$ and the area of the junction to be $10^{-6} \, \text{m}^2$. (Hint: find the charge in the diode at $-10 \, \text{V}$ bias.)

3.4. Design a diode which passes 10 mA when a forward bias of $0 \cdot 5$ V is applied. (Note: diffusion distances can be controlled in manufacture—a value in the range 10^{-5}–$10^{-4} \, \text{m}$ would be reasonable.)

3.5. Show that in a symmetrical abrupt junction, the peak field $= \frac{1}{2}(V_j/x)$, where x is the total thickness of the depletion layer, and V_j is the total potential across the junction. Does the relation still hold if the junction is not symmetrical, or not abrupt?

3.6. A reverse-biased p–n junction diode has a capacitance of 10 pF. Calculate the total depletion layer width, and the difference in potential between the p- and n-regions, given that the area of the junction is $1 \cdot 0 \times 10^{-7} \, \text{m}^2$, $\varepsilon_r \times \varepsilon_0 = 1 \cdot 0 \times 10^{-10} \, \text{F m}^{-1}$, and that $N_a = N_d = 10 \times 10^{22} \, \text{m}^{-3}$.

3.7. A p–n junction has $N_a = 10^{22} \, \text{m}^{-3}$ and $N_d = 10^{24} \, \text{m}^{-3}$. Find hole and electron densities at the edges of the depletion layer with a forward bias of $0 \cdot 5$ V. Take $n_i = 10^{16} \, \text{m}^{-3}$.

4. Junction transistors and thyristors

Bipolar transistors

The bipolar junction transistor, which rapidly supplanted the earlier point-contact transistor, has been the financial foundation of semiconductor electronics. From now on, when the term 'transistor' is used without qualification, then a bipolar junction transistor should be understood.

Construction and low-frequency characteristics

A simple version of the shape of the active regions of a transistor is shown in Fig. 4.1(a). The three layers of semiconductor which constitute the emitter, base, and collector may be n–p–n or p–n–p, giving two alternative arrangements. The two kinds of transistor complement each other's function in electronic circuits, the signs of currents and voltages being reversed. The

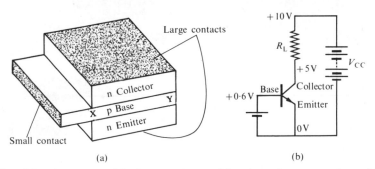

Fig. 4.1. An n–p–n bipolar transistor: (a) schematic construction; (b) conventional symbol and circuit showing 'normal bias'. V_{cc} is the supply to the collector and load.

physical explanations required for the two kinds are the same if holes be read for electrons, p for n, and so on. The main carrier flow is from the emitter through the base and out through the collector, so the emitter and collector have large contacts. The base is thin—as thin as can be made ($0\cdot2\,\mu$m or less), and it has a contact at the side for a third controlling connection.

In use the base-emitter junction is forward-biased (as if it were a diode) and the base-collector junction is reverse-biased by 1–20 V (Fig. 4.1(b)). Figure 4.2(a) and (b) show the collector

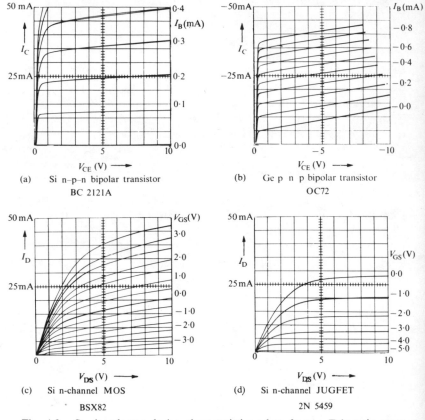

(a)   Si n–p–n bipolar transistor
BC 2121A

(b)   Ge p n p bipolar transistor
OC72

(c)   Si n-channel MOS
BSX82

(d)   Si n-channel JUGFET
2N 5459

Fig. 4.2. Semiconductor device characteristics taken from a Tektronix curve tracer; (a) the wide spacing of the curves show that dI_C/dI_B is large; (b) the p–n–p device has the sign of I_C and V_{CE} reversed; the tips of the curves mark out a 'load line'; (c) V_{GS} can be positive or negative; the traces near the origin show some instrumental kinks; (d) the JUGFET is more closely related to the MOS than to the bipolar transistor.

current as a function of collector-emitter voltage, for various values of the base current for a Si n–p–n transistor and a Ge p–n–p transistor. Notice that the current increases very rapidly as V_{CE} goes from 0 V to about 0·5 V, and then increases much less rapidly. For the Si transistor, the collector current is very close to zero when the base current is zero, but this is not so for the Ge transistor. An important parameter is (dI_C/dI_B) for constant V_{CE} (often called β). The steps of I_B in the two sets of curves are the same, so the larger spacing of the curves for the Si device implies a higher β (what is its numerical value?).

With a suitable load resistor R_L the collector voltage can vary by much more than the changes in the base-emitter voltage. Consequently small changes of current and voltage can be amplified into much larger ones, a function with many applications.

The limitations of a transistor in amplifying high-frequency signals cannot be assessed from curves such as those in Fig. 4.2. The high-frequency characteristics are discussed on p. 111.

The maximum current that can be passed through a transistor is limited by the need to keep the transistor from overheating. High-power transistors need good heat sinks. The maximum voltage that can be applied between collector and base may be limited by avalanche breakdown in the collector-base depletion layer, or by the depletion layer extending right across the base and reaching the emitter, this process being called 'punch through'. Maximum values of V_{CE} may be from 5 to 1500 V.

Physical principles of bipolar transistors

Three physical principles make a useful first description of the way a bipolar transistor works.

(a) The externally applied voltages control the ratio of the carrier densities on the two sides of each depletion layer (see p. 62).

(b) Away from the depletion layer the regions are neutral.

(c) Hence carriers move away from the junctions after injection by diffusion and the action of any fields built in by variable doping. The applied voltages control the carrier flow solely through the injection process.

Figure 4.3 shows these ideas for a p–n–p transistor from three aspects: the variation along a line from emitter to collector of the potential and the band structure, the fluxes of particles, and the carrier densities.

In Fig. 4.3(a), the width that a stream is drawn indicates the current it carries, and the arrows point in the direction of particle flow not in the direction of the conventional electrical current. The largest current is of holes injected into the base. A fraction f_1 of these holes combine with electrons in the base, but most diffuse to the collector-base depletion layer where they are whisked away into the collector by the high field in the depletion layer.

Injection always works both ways, though to different extents, and there is a current of electrons injected back into the emitter, which is proportional to the forward current of holes I_{Eh} (let it be $f_2 I_{Eh}$). In useful transistors f_1 and f_2 are arranged to be small.

The total emitter current is $I_{Eh}(1 + f_2)$, while the total base current is $I_{Eh}(f_1 + f_2)$, so their ratio is

$$(1 + f_2)/(f_1 + f_2) \sim 1/(f_1 + f_2).$$

This is the common-emitter current gain β of the transistor.

From the collector there is a fairly constant leakage current to the base of electrons. In Si transistors this is often so small that it can be ignored.

The mention of both electrons and holes in this discussion justifies the name *bipolar* transistor, by comparison with some field-effect transistors, (see p. 138) where only the majority carriers appear to play an important part. In Fig. 4.3(b), the relative positions of the Fermi level show the external bias voltages. In a p–n–p transistor the base is slightly negative relative to the emitter, and the collector is very negative. There is no field in the base, which is correct for a uniformly doped base. The results of non-uniform doping in the base are discussed on p. 112. Notice how the majority carriers in the base are in a potential well, with a barrier on each side of them. This explains why the minority carriers carry most of the current, because the majority carriers are, to a useful extent, trapped in the base.

The depletion region between emitter and base is thin because the potential across it is small. The depletion region between base and collector has a much higher voltage across it, and its thickness has to be taken into account (see p. 62).

(a)

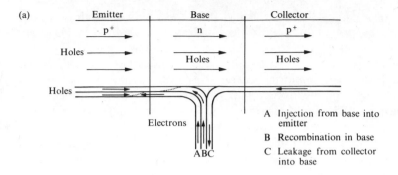

A Injection from base into
   emitter

B Recombination in base

C Leakage from collector
   into base

(b)

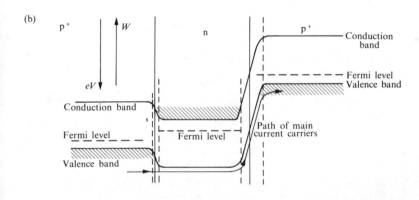

(c)

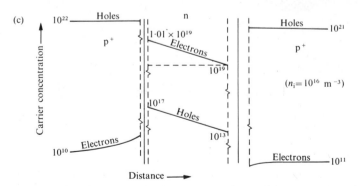

Fig. 4.3.   Processes in a bipolar transistor; (a) fluxes of holes and electrons; (b) band structure and potentials; (c) hole and electron concentrations.

The carrier densities in Fig. 4.3(c) are difficult to draw so that all the points are clearly demonstrated, as both very large and very small densities are important. In the base the density of the majority carriers (electrons) makes a line of the same slope as that for the minority carriers, so the region is everywhere neutral. The electrons ought therefore to diffuse, but since they are trapped, a small field builds up which is just sufficient to balance the diffusion current, but, except at high currents, is too small to have a noticeable effect on the far fewer holes. The injection of carriers back into the emitter can be seen, but the recombination of carriers in the base is too small an effect to be evident, though if it were shown the density in the base would fall a little more steeply near the emitter than near the collector.

The Ebers–Moll model for a bipolar transistor

The ideas sketched in Fig. 4.3 can be put down in equations, and we shall do this for a simple case where all three regions are so short that recombination can be completely neglected except at the contacts (Fig. 4.4). Our results will apply to any combination of bias voltages—large or small, positive or negative—and lead to the Ebers–Moll equations for a bipolar transistor, which describe d.c. behaviour in a very compact way. Were we to take recombination in the bulk into account, the mathematics would become more involved, but the form of the final equation would be unaltered.

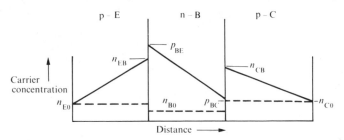

Fig. 4.4. Model for the Ebers–Moll analysis. Recombination occurs only at contacts, so the graphs of n or p against distance are straight. Read a symbol like n_{CB} as 'the concentration of electrons at the side of the collector nearer the base', and a symbol like p_{B0} as 'the concentration of holes in the emitter in equilibrium'.

The first of the three principles stated at the beginning of the previous section may be expressed using the notation of Fig. 4.4, for the base-emitter junction,

$$\frac{p_{BE}}{p_{B0}} = \exp\left(\frac{eV_{EB}}{\kappa T}\right) = \frac{n_{EB}}{n_{E0}}, \tag{4.1}$$

and for the base-collector junction,

$$\frac{p_{BC}}{p_{B0}} = \exp\left(\frac{eV_{CB}}{\kappa T}\right) = \frac{n_{CB}}{n_{C0}}, \tag{4.2}$$

For a transistor of area A without any field built into the base, the second and third principles allow us to describe the total current as made up of diffusion currents.

For the base, the hole current is

$$I_{Bp} = AeD_h(p_{BE} - p_{BC})/W_B. \tag{4.3}$$

For the emitter and collector, the electron-diffusion current is

$$I_{En} = AeD_e(n_{EB} - n_{E0})/W_E$$
$$I_{Cn} = AeD_e(n_{CB} - n_{C0})/W_C. \tag{4.4}$$

Because volume recombination is absent from our model, the hole current in the base (eqn (4.3)) must also be the hole current in the emitter and collector, so we know the total current in each region. The total base current is fixed by

$$I_E + I_B + I_C = 0. \tag{4.5}$$

where we are defining a positive current as flowing into each terminal. This is the end of the physics—the rest is algebra to put the equations into revealing forms.

It is useful to define three quantities which are characteristic of the three parts of the transistor,

$$E = AeD_e n_{E0}/W_E,$$
$$B = AeD_h p_{B0}/W_B,$$
$$C = AeD_e n_{C0}/W_C.$$

The total emitter and collector currents, I_E and I_C, may now be written out, using (4.1) and (4.2) for the densities in (4.3) and

(4.4),

$$I_E = I_{En} + I_{Ep} = E\left\{\exp\left(\frac{eV_{EB}}{\kappa T}\right) - 1\right\}$$

$$+ B\left\{\exp\left(\frac{eV_{EB}}{\kappa T}\right) - \exp\left(\frac{eV_{CB}}{\kappa T}\right)\right\}$$

$$I_C = I_{Cn} + I_{Cp} = C\left\{\exp\left(\frac{eV_{CB}}{\kappa T}\right) - 1\right\} \qquad (4.6)$$

$$+ B\left\{\exp\left(\frac{eV_{CB}}{\kappa T}\right) - \exp\left(\frac{eV_{EB}}{\kappa T}\right)\right\}.$$

In these equations the transistor looks like a voltage-controlled device, because if the voltages are known the currents can easily be found. This approach is helpful in understanding why the transistor works, but a transformation can put the equations into another form which has been found more suitable for circuit work.

First we re-cast each expression in (4.6) into the form of diode equations:

$$I_E = (E + B)\left\{\exp\left(\frac{eV_{EB}}{\kappa T}\right) - 1\right\}$$

$$- \left(\frac{B}{C + B}\right)(C + B)\left\{\exp\left(\frac{eV_{CB}}{\kappa T}\right) - 1\right\}$$

$$I_C = (C + B)\left\{\exp\left(\frac{eV_{CB}}{\kappa T}\right) - 1\right\} \qquad (4.7)$$

$$- \left(\frac{B}{E + B}\right)(E + B)\left\{\exp\left(\frac{eV_{EB}}{\kappa T}\right) - 1\right\}.$$

Next we introduce symbols which relate to directly measureable properties of the transistor in place of E, B, and C,

$$I_{EB0} = (E + B), \qquad I_{CB0} = (C + B),$$

$$\alpha_i = B/(C + B), \qquad \alpha_n = B/(E + B).$$

Clearly, $\alpha_n I_{EB0} = \alpha_i I_{CB0}$. The Ebers–Moll equations now take the form

$$I_E = I_{EB0} \left\{ \exp\left(\frac{eV_{EB}}{\kappa T}\right) - 1 \right\} - \alpha_i I_{CB0} \left\{ \exp\left(\frac{eV_{CB}}{\kappa T}\right) - 1 \right\};$$

$$I_C = I_{CB0} \left\{ \exp\left(\frac{eV_{CB}}{\kappa T}\right) - 1 \right\} - \alpha_n I_{EB0} \left\{ \exp\left(\frac{eV_{EB}}{\kappa T}\right) - 1 \right\}. \tag{4.8}$$

To obtain a value for I_{EB0}, we measure I_E when the emitter-base junction is reverse-biased so that $\exp(eV_{EB}/\kappa T)$ is very small, and with the collector open circuit, i.e. with $I_C = 0$ (check the relation between I_{EB0} and the measured value of I_E—they are not the same). A value of I_{CB0} can be found in an equivalent way. To find α_n (α_{normal}) we make V_{CB} very large and negative, so that

$$I_C = I_{CB0} - \alpha_n I_{EB0}\{\exp(eV_{EB}/\kappa T) - 1\} = I_{CB0} - \alpha_n(I_E - \alpha_i I_{CB0}).$$

Then $\alpha_n = -dI_C/dI_E$. The bias condition is the normal one for the ordinary use of the transistor, and we expect α_n to be a little less than unity. If the functions of the emitter and collector are interchanged, the transistor will still work as a transistor, though probably less well. The value of α_i ($\alpha_{inverted}$) can be found by making V_{EB} large and negative, and measuring $-dI_E/dI_C$.

We can now build up an equivalent circuit for the d.c. operation of a transistor from eqn (4.8). The term $I_{EB0}\{\exp(eV_{EB}/\kappa T) - 1\}$ could be the description of a junction diode with $I_r = I_{EB0}$. We can represent it on the equivalent circuit by a diode symbol, with I_{EB0} written beside it. The diode goes between emitter and base, the current is part of the emitter current, and the diode voltage is V_{EB}.

The term $-\alpha_i I_{CB0}\{\exp(eV_{CB}/\kappa T) - 1\}$ requires a current generator which states that the emitter current has a contribution $-\alpha_i I_C^*$ (I_C^* is defined in Fig. 4.5(a)). The components in the collector branch are read in the same way from the equation for I_C. The process of building up Fig. 4.5(a) by reading the terms in the equations should be noted, as it is a technique characteristic of the physics of electronic devices.

When the normal set of bias voltages are applied to a transistor (V_{CB} large and negative, V_{EB} small and positive for our p–n–p transistor), Fig. 4.5(a) simplifies to Fig. 4.5(b), if we may neglect

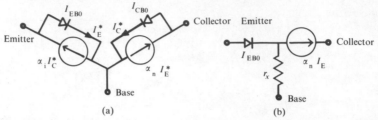

Fig. 4.5. Ebers–Moll equivalent circuit: (a) for any bias condition: (b) a simplified circuit which applies when normal working bias is applied, and r_x is added.

I_{CB0}. The resistor in the base branch is not part of the formal Ebers–Moll model, but has been added to the circuit to allow for the resistance of the semiconductor away from the active region of the base. We shall use Fig. 4.5(b) as the basis for our analysis of the noise properties of a bipolar transistor, and also show how it can lead to the hybrid-π small-signal model for a bipolar transistor.

The Ebers–Moll equivalent circuit applies for all bias conditions. If the standard conditions are applied of forward bias across the base-emitter junction and reverse bias across the base-collector junction, then, for a silicon device, the current through the collector diode is negligible, and so therefore is the current in the current generator in the emitter branch. The

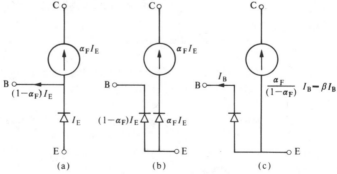

Fig. 4.6. Equivalent circuit transformation. The Ebers–Moll model may be simplified in (a) when the collector-base junction is reverse-biased. It can then be transformed until it is one step away from the hybrid-π model. The final step (not shown in this figure) is to replace the diode in (c) by its small-signal representation.

simpler equivalent circuit of Fig. 4.6(a) is then obtained. This can be transformed into Fig. 4.6(b) and then to Fig. 4.6(c), which is very close to the form used in the hybrid-π equivalent circuit in its basic form. The hybrid-π equivalent circuit considers only small changes away from a bias point, but includes frequency-dependent elements, and is fully discussed later in this chapter.

The Ebers–Moll equations go a long way to giving a good description of the d.c. characteristics of bipolar transistors, but they omit an important effect. As V_{CB} increases the collector-base depletion layer expands, so that the remaining neutral base width becomes less and less. As a result the collector current increases with V_{CE}, as can be seen in Fig. 4.2(a) and (b). If the current were constant the slope resistance would be infinite, and would lead to very different circuit effects, so the change of base width or base-width modulation is important.

Recombination effects

This section will present an analysis of a bipolar transistor which goes one step deeper than the previous section. Injection of electrons from base to emitter will be included, as before, but now recombination in the emitter, base, and collector will be considered. The analysis will start with statements which apply to any bias, but will then consider only the bias which is applied in the normal active mode of operation.

Once again we shall consider a p–n–p transistor. The continuity equation for minority carriers is the starting point. In the base it is

$$\frac{\mathrm{d}p}{\mathrm{d}t} = -\frac{p - p_{B0}}{\tau_B} + D_B \frac{\mathrm{d}^2 p}{\mathrm{d}x^2},$$

where p is the hole concentration at x, p_{B0} is the equilibrium value of p as in Fig. 4.4, and x is measured from the emitter side of the base region. τ_B is the recombination time for minority carriers in the base. Equivalent equations apply to the minority carriers in the collector and emitter.

The boundary conditions for these differential equations are set by the applied voltages, which control the minority carrier concentrations at the edge of each depletion layer.

Thus, at the emitter-base depletion layer

$$p'(0) = p(0) - p_{B0}$$
$$= p_{B0}[\exp(eV_{EB}/\kappa T) - 1]$$
$$n'(x_E) = n(x_E) - n_{E0}$$
$$= n_E[\exp(eV_{EB}(\kappa T) - 1],$$

and at the collector-base depletion layer

$$p'(W) = p(W) - p_{B0}$$
$$= p_{B0}[\exp(eV_{CB}/\kappa T) - 1]$$
$$n'(x_C) = n(x_C) - n_{C0}$$
$$= n_{C0}[\exp(eV_{CB}/\kappa T) - 1].$$

Here $p'(x)$ and $n'(x)$ are the excess minority-carrier concentrations at x, as shown in Fig. 4.7.

A general solution to the differential equations for n' and p' can be written down. A simpler version applies when emitter and collector are much thicker than L_E and L_C, respectively. We take this case, and consider the normal active bias mode conditions, where the collector-base junction is reverse-biased. The value of $\exp(eV_{CB}/\kappa T)$ is then very small, and the solution for the

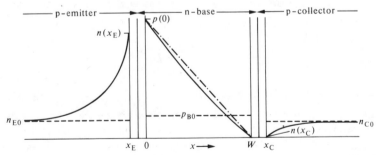

Fig. 4.7. The minority-carrier densities for a bipolar transistor with the emitter-base junction forward-biased, and the collector-base junction reverse-biased. In all three parts of the transistor the graph of minority-carrier density is curved, showing that recombination is taking place.

minority carriers becomes

$$p(x) = p_{B0} + p'(0)\frac{\left[\exp\dfrac{W}{L_B}\exp\left(-\dfrac{x}{L_B}\right) - \exp\left(-\dfrac{W}{L_B}\right)\exp\left(\dfrac{x}{L_B}\right)\right]}{2\sinh(W/L_B)}$$

$$n(x) = n_{E0} + n'(x_E)\exp - [-(x - x_E)/L_E] \tag{4.9}$$

$$n(x) = n_{C0} + n'(x_C)\exp[-(x - x_C)/L_C] \tag{4.10}$$

in base, emitter, and collector. $L_{B,E,C}$ are the diffusion lengths in base, emitter, and collector, so that

$$L_B = (\tau_B D_B)^{\frac{1}{2}} \text{ etc.}$$

The solutions for collector and emitter are clear. The solution for the base can be put in a simpler form for the usual situation where $W \ll L_B$. In this case x in the base will also satisfy $x \ll L_B$, so that series expansions for $\exp(\pm W/L_B)$ and $\exp(\pm x/L_B)$ are possible. It is necessary to take the first three terms of each series and, if this is done,

$$p(x) = p_{B0} + p'(0)\left[\underbrace{1 - \frac{x}{W}}_{\substack{\text{Linear gradient}\\\text{leads to}\\\text{diffusion current}}} + \underbrace{\frac{x^2}{2L_B^2} - \frac{xW}{2L_B^2}}_{\substack{\text{recombination}\\\text{in base}}}\right]. \tag{4.11}$$

The next step is to find the terminal currents. This is done by adding the minority current arriving at the two sides of a depletion layer. Thus,

$$I_E = -AeD_B\frac{dp}{dx}\bigg|_{x=0} - AeD_E\frac{dn}{dx}\bigg|_{x=x_E}.$$

Equations 4.11 and 4.9 allow the evaluation of dp/dx and dn/dx.

$$\frac{dp(x)}{dx} = p'(0)\left\{-\frac{1}{W} + \frac{x}{L_B^2} - \frac{W}{2L_B^2}\right\}$$

and hence, at $x = 0$,

$$\frac{dp(x)}{dx}\bigg|_{x=0} = p'(0)\left\{-\frac{1}{W} - \frac{W}{2L_B^2}\right\}$$

and at $x = x_E$,

$$\frac{dn(x)}{dx}\bigg|_{x=x_E} = n'(x_E)/L_E.$$

The formula for the emitter current is then

$$I_E = AeD_Bp_{B0}\left(\exp\left(\frac{eV_{EB}}{\kappa T}\right) - 1\right)\left\{\frac{1}{W} + \frac{W}{2L_B^2}\right\}$$
$$+ AeD_En_{E0}\left(\exp\left(\frac{eV_{EB}}{\kappa T}\right) - 1\right)\frac{1}{L_E}.$$

The collector current is found in the same way, taking the total of the two minority-carrier diffusion currents at the boundaries of the collector-base depletion layer

$$I_C = AeD_Bp_{B0}\left(\exp\left(\frac{eV_{EB}}{\kappa T}\right) - 1\right)\left\{\frac{1}{W} - \frac{W}{2L_B^2}\right\}$$
$$+ AeD_Cn_{C0}(-1)/L_C.$$

The base current I_B is now found from

$$I_C + I_E + I_B = 0$$

and is

$$I_B = -Ae\left\{\underbrace{D_Bp_{B0}\frac{W}{L_B^2}}_{\substack{\text{recombination} \\ \text{in base}}} + \underbrace{\frac{D_En_{E0}}{L_E}}_{\substack{\text{injection into} \\ \text{emitter}}}\right\}\left(\exp\left(\frac{eV_{EB}}{\kappa T}\right) - 1\right) + \underbrace{\frac{AeD_Cn_{C0}}{L_C}}_{\text{leakage from collector}}.$$

$$(4.12)$$

Equation (4.12) is a main result of the analysis, as it shows how the construction of the device leads to a control of an important property, the base current. The small signal-current gain β_0 can now be derived

$$\beta_0 = \frac{dI_C}{dI_B} = \frac{dI_C}{dV_{EB}}\bigg/\frac{dI_B}{dV_{EB}}$$

$$= \frac{AeD_Bp_{B0}\left\{\dfrac{1}{W} - \dfrac{W}{2L_B^2}\right\}\exp\left(\dfrac{V_{EB}}{\kappa T}\right)\cdot\dfrac{e}{\kappa T}}{Ae\left\{\dfrac{Wp_{B0}D_B}{L_B^2} + \dfrac{n_{E0}D_E}{L_E}\right\}\exp\left(\dfrac{V_{EB}}{\kappa T}\right)\dfrac{e}{\kappa T}}$$

Hence, since $W/2L_B^2 \ll 1/W$,

$$\beta_0 = \cfrac{1}{\cfrac{W^2}{L_B^2} + \cfrac{n_{E0}}{p_{B0}}\cfrac{D_E}{D_B}\cfrac{W}{L_E}} \qquad (4.13)$$

The conditions for a high value of β_0 can now be seen. Both of the two terms in the bottom of the fraction in eqn (4.13) must be small. Therefore $W \ll L_B$ is essential, and since D_E and D_B are not very different, it may be useful to have $n_{E0} \gg p_{B0}$.

In modern transistors W is so small that it may not be necessary to require that n_{E0} is much greater than p_{B0}.

Example. Take $W = 1 \times 10^{-6}\,\text{m}$; $L_{B,E} = 4 \times 10^{-5}\,\text{m}$; $N_{AE} = 10^{24}\,\text{m}^{-3}$; $N_{DB} = 10^{23}\,\text{m}^{-3}$; and $D_E/D_B = 3$. Then,

$$\frac{W}{L_{B,E}} = \frac{1}{40} \text{ and } \frac{n_E}{p_B} = \frac{1}{10}.$$

Hence,

$$\beta_0 = \cfrac{1}{\left(\cfrac{1}{40}\right)^2 + \cfrac{1}{10} \times 3 \times \cfrac{1}{40}} = 123.$$

The major term in this example is injection of minority carriers into the emitter, with recombination in the base being a minor factor.

Hybrid-π equivalent circuit for bipolar transistors

Transistors are often used to amplify small signals, so it is important to have some model to describe the way a transistor operates in an amplifying circuit. We assume that the transistor has suitable steady bias voltages and currents, and shall expect the values of the components of the equivalent circuit to depend on the quiescent conditions. We wish our model to apply over a wide range of frequency with reasonable accuracy.

Many small-signal equivalent circuits have been suggested for transistors, and indeed they can be transformed one into another. Here we examine the hybrid-π model, because its elements can be related to the physical processes occurring in the

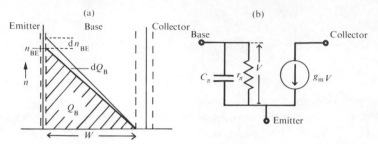

Fig. 4.8. Hybrid-π small-signal equivalent circuit: (a) the excess charge in the base is Q_B, and the density gradient $dn/dx = n_{BE}/W$: (b) the first three elements of the hybrid-π equivalent circuit.

transistor. We start with the idea that the density gradient of excess carriers in the base is constant (or nearly so), and that the current is proportional to the density gradient (Fig. 4.8(a)),

$$I_C = -AD_e e(dn/dx).$$

The idea holds if the carriers have time to distribute themselves properly—when this condition is not met, the frequency of operation is so high that the gain of the transistor has become low.

The density falls from $n_{BE} = n_{B0}\exp(eV_{EB}/\kappa T)$ on the emitter side of the base to a low value on the collector side, so the density gradient is

$$\frac{dn}{dx} = -(n_{B0}/W)\exp(eV_{EB}/\kappa T),$$

where we use the same notation as in Fig. 4.4. Thus

$$I_C = (AeD_e n_{B0}/W)\exp(eV_{EB}/\kappa T).$$

$$\left.\frac{dI_C}{dV_{EB}}\right|_{V_{CE}} = \frac{e}{\kappa T}I_C$$

or

$$dI_C = (e/\kappa T)I_C\, dV_{EB}. \tag{4.14}$$

Equation (4.14) can be put in terms of a transfer conductance $g_m = (e/\kappa T)I_C$, as we have a change in the voltage between two terminals controlling a current to the third terminal.

Then $dI_C = g_m \, dV_{EB}$, and we have the first element of the hybrid-π equivalent circuit (Fig. 4.8(b)). A particular convenience is that the value of g_m depends only on quantities that are easy to find. The steady collector current is related to the total excess charge in the base Q_B,

$$I_C = eAD_e n_{BE}/W,$$
$$Q_B = \tfrac{1}{2} eA n_{BE} W. \tag{4.15}$$

Thus

$$I_C = \frac{Q_B}{(W^2/2D_e)}.$$

Since current is a rate of flow of charge, $W^2/2D_e$ is the time taken for Q_B to flow through the base, and is given the symbol τ_t, the minority-carrier transit time.

The change in the base current when the base-emitter voltage is altered from its quiescent value has two components—steady and transient. The steady part supplies the change in the loss processes in the base, namely, the recombination of carriers in the base and the injection of carriers back into the emitter. If the excess majority-carrier lifetime in the base due to these causes is τ_B, the steady current needed to replenish the carrier density is

$$I_B = Q_B/\tau_B.$$

Since Q_B is related to n_{BE} (eqn (3.33)) and hence to V_{EB}, we can find how I_B is related to V_{EB},

$$\frac{dI_B}{dV_{EB}} = \frac{eAW}{2\tau_B} \cdot \frac{d}{dV_{EB}} \left\{ n_{B0} \exp\left(\frac{eV_{KB}}{\kappa T}\right) \right\}.$$

Thus we now have an expression for g_π, the conductance between base and emitter,

$$g_\pi = \frac{e^2 AW}{2\tau_B} \cdot \frac{n_{B0}}{\kappa T} \exp\left(\frac{eV_{KB}}{\kappa T}\right) = \frac{1}{r_\pi}, \tag{4.16}$$

where r_π, the resistance between base and emitter, is used as an alternative notation to $1/g_\pi$.

A little substitution (try it) shows that

$$g_\pi = g_m \cdot \tau_t/\tau_B. \tag{4.17}$$

Since τ_t is always much less than τ_B, $g_\pi \ll g_m$, and only small changes in I_B are needed to produce large changes in I_C. The ratio $(\tau_t/\tau_B)^{-1}$ is β, the common-emitter current gain.

The transient part of the base current is needed to bring Q_B up to its new value

$$\frac{dQ_B}{dV_{EB}} = \frac{dn_{BE}}{dV_{BE}} \cdot \frac{eAW}{2} = \frac{e}{\kappa T} Q_B = \frac{e}{\kappa T}(I_C\tau_t).$$

The expression dQ_B/dV_{EB} describes a capacitance C_π, in the equivalent circuit, where

$$C_\pi = (e/\kappa T)(I_C\tau_t) = g_m\tau_t.$$

In Fig. 4.8(b) we have the three main elements of the hybrid-π equivalent circuit, and these alone would give a useful approximate description of the variation of the performance of the transistor with frequency.

The next group of components depends on base-width modulation. This is the reduction of W as V_{CB} increases (Fig. 4.9(a)), which has three important consequences.

There is an increase in dn/dx, and hence I_C for a given V_{BE}. On the equivalent circuit we represent this by a conductance g_0 between emitter and collector. A value for g_0 is best obtained from curves like those in Fig. 4.2, rather than by calculation (but see problem 4.4).

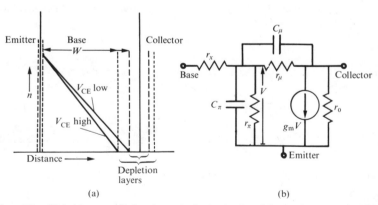

(a)   (b)

Fig. 4.9. Hybrid-π small-signal equivalent circuit: (a) an increase in the collector-emitter voltage difference reduces the base width and increases dn/dx: (b) the complete hybrid-π equivalent circuit.

Another result of base width modulation is that Q_B is reduced, so a transient reverse current is called for. A capacitance C_μ between collector and base provides this. The reduced Q_B requires reduced replenishment, so the total steady base current should reduce as V_{CE} increases. A conductance g_μ between collector and base does this job. Notice that C_μ and g_μ are the first components we have discussed that allow the feedback of a signal from collector to base. In general, the smaller they are, the better. Like g_0, g_μ and C_μ are better determined from measurement rather than calculation.

The two capacitances C_μ and C_π are physically related to the diffusion capacitance analysed on p. 72. The other kind of p–n junction capacitance—depletion-layer capacitance—is also present at the two junctions. In the equivalent circuit the depletion-layer capacitances may be added on to C_μ and C_π.

Current flowing to the active region of the base between collector and emitter has first to pass through a volume of lightly doped semiconductor, which may have enough resistance to matter. It is found to be worthwhile including this on the equivalent circuit as an extra resistance r_x, which is independent of the bias conditions, having a value between 10 and 100 Ω. The voltage which controls the current generator g_m is now not the voltage between the emitter and base terminals, but the voltage relative to the emitter of an internal node on the equivalent circuit, which represents the signal reaching the p–n junction.

The full hybrid-π equivalent circuit (Fig. 4.9(b)) should give a close description of the way a transistor with one set of bias conditions handles small signals over its full range of working frequencies.

High-frequency limits for a bipolar transistor

The amplification available from a transistor circuit falls at high frequencies. A useful figure of merit is the frequency (f_t) where the current gain in the common-emitter circuit with zero load impedance has fallen to unity. The value of f_t can be related to the total delay of minority carriers passing from emitter to collector. Delays which may be important can occur in changing the charge at the emitter-base depletion layer, in diffusing across the base, and in changing the charge at the collector.

In some transistors the delay τ_t in diffusing across the base, is the longest. We have seen that $\tau_t = W^2/2D$, and hence expect

$$f_t = 1/(2\pi\tau_t) = D/(\pi W^2).$$

This conclusion can be confirmed by looking at a simple version of the hybrid-π equivalent circuit which omits C_μ and g_μ. The only frequency-dependent element is C_π, which in parallel with g_π gives a time constant τ_t (check this using the expressions for C_π and g_π—remember that the time constant of a capacitance and a conductance is C/g).

For transistors made by the planar process there can be a field in the base assisting the motion of the minority carriers. If the doping density at the emitter and collector sides of the base are N_{BE} and N_{BC} then the difference between Fermi level and conduction band is

$$(W_C - W_F)_E = \frac{\kappa T}{e} \ln(N_{BE}/n_i)$$

and

$$(W_C - W_F)_C = \frac{\kappa T}{e} \ln(N_{BC}/n_i)$$

for the two boundaries of the base. As a result the difference of potential across the base is

$$\frac{\kappa T}{e} \ln(N_{BE}/N_{BC}).$$

An exponential variation of the doping density across the base leads to the potential difference being distributed as a uniform field across the base. In such a case the drift time across the base is

$$\tau_{dr} = \frac{W}{\mu E} = \frac{W^2}{\mu V} = eW^2/(\mu\kappa T \ln(N_{BE}/N_{BC}))$$

and hence

$$\tau_{dr} = W^2/D \ln(N_{BE}/N_{BC}).$$

Thus the ratio of transit time across a base by diffusion and by drift is

$$(\tfrac{1}{2})/\ln(N_{BE}/N_{BC}).$$

Values of N_{BE} of 10^{23} and N_{BC} of 10^{20} are about the maximum range, so that the transit of holes may be speeded up by about a factor of 4 using this process.

Transistors are available with values of f_t up to 10 GHz; at these frequencies the connecting leads have to be designed to be part of the microwave circuit—merely making them short is not enough.

The charge-control model

The charge-control model for bipolar transistors is used to describe their action in pulse circuits. Unlike the hybrid-π equivalent circuit, it does not describe small variations about a specified bias point, but is suitable for situations where the transistor changes from being completely on to completely off, as in many digital applications.

In the charge-control model, attention is focused on the excess charge Q_B in the base of the transistor, and the equivalent circuit is built up of components whose value is controlled by Q_B. The need to build up and replenish Q_B is expressed by some of the equivalent-circuit components.

On p. 109 we saw that the relations between Q_B, I_C, and I_B were

$$I_C = Q_B/\tau_t,$$
$$I_B = Q_B/\tau_B. \tag{4.18}$$

These relations will apply both to transistors with a uniformly doped base, when $\tau_t = W^2/2D$, and to transistors with a density gradient of doping atoms in the base. If we write $\tau_B = \beta\tau_t$ we can see equivalent circuit components corresponding to eqn (4.18) in Fig. 4.10(a). The symbol labelled Q_B is not a capacitor, as the voltage across it does not increase as the charge increases; it is sometimes called a *storance*. The full set of equations corresponding to Fig. 4.10(a) is

$$I_B = Q_B/\beta\tau_t + dQ_B/dt,$$
$$I_C = Q_B/\tau_t, \tag{4.19}$$
$$I_E + I_C + I_B = 0.$$

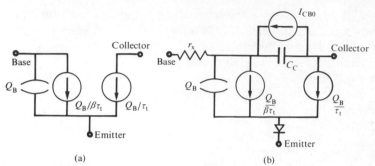

Fig. 4.10. Charge control large-signal equivalent circuit: (a) the three circuit elements shown form the simplest charge control equivalent circuit; Q_B is the excess charge (majority or minority) stored in the base: (b) an extended equivalent circuit.

This is the basic charge-control model, and describes both a.c. and d.c.; but to be useful it needs further development.

Figure 4.10(b) is a full model for conditions where the transistor is not saturated, i.e. where the collector-base junction is reverse-biased, and all minority carriers reaching it are swept across the depletion layer. The new components of the equivalent circuit are:

(a) A current generator representing the leakage current from collector to base, taken to be independent of V_{CB}.

(b) The depletion-layer capacitance C_C for the collector-base junction. To make the model simple we take an average value for C_C, even though as V_{CB} increases, C_C will fall.

(c) The depletion-layer capacitance C_E for the emitter-base junction. We can take an average value, or noting that V_{BE} varies only little, realize that the charge in C_E is unchanging and ignore it.

(d) A diode in the emitter lead. It represents the voltage between emitter and base, and can be as exact a description of a diode as is required, but has no stored charge or delays—it is an 'instantaneous diode'.

(e) A resistance r_x, representing the passive base region.

The equations corresponding to Fig. 4.10(b) are

$$I_B = Q_B/\beta\tau_t + dQ_B/dt + dQ(C_C)/dt - I_{CB0} = 0,$$

$$I_C = Q_B/\tau_t - dQ(C_C)/dt + I_{CB0} = 0, \qquad (4.20)$$

$$dQ(C_C)/dt = -C_C \, dV_{CB}/dt.$$

Notice that C_E has been ignored, and r_x and the diode appear only when the terminal voltages are being calculated.

When the base current is larger than is needed to maintain the maximum current that the circuit can attain (V_{CC}/R_L Fig. 4.1(b)), the current and the base charge are no longer related by $I_C = Q_B/\tau_t$; a new relation is needed. The base charge has to be divided between Q_B which still relates to I_C as before, and Q_{Bs}, the saturated base charge (Fig. 4.11(a)), which does not contribute to the density gradient though it is stored in the base. The saturated base charge decays at a rate $dQ_{Bs}/dt = Q_{Bs}/\tau_s$, where τ_s is comparable to $\beta\tau_r$, though it may well be somewhat shorter.

If, in a circuit, V_{CC} were suddenly to increase or R_L were to decrease, then Q_{Bs} could be redistributed to permit a rapid increase in I_C without the need for an increase in the base current. Another situation in which Q_{Bs} has to be considered occurs when I_B falls to zero after a pulse. Until all the charge in the base has decayed (Q_{Bs} as well as Q_B) collector current can continue to flow. In practice this means that a transistor takes far longer to switch off than would otherwise be predicted.

The equations describing the charge-control model (Fig. 4.11) for saturated conditions are

$$I_B = Q_B/\beta\tau_t + Q_{Bs}/\tau_s + dQ_{Bs}/dt,$$
$$I_C = Q_B/\tau_t. \tag{4.21}$$

The charge-control model is used by defining a starting situation and calculating how long a transistor takes to reach successive stages through a pulse, employing judicious approximations at each stage.

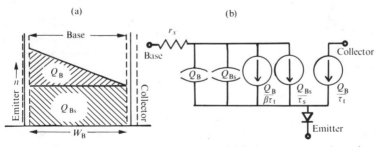

Fig. 4.11. The saturated charge-control model: (a) the base charge is made up of two parts, Q_B and Q_{Bs}, (b) the equivalent circuit is used while $I_B > I_C/\beta$.

Two simplifications in the charge-control model presented here are the omission of any explicit description of base-width modulation (some average base width must be assumed) and the assumption that the density gradient is linear. The effect of a graded base doping introduces no error when the right values of τ_t are used, but a brief time (about $\frac{1}{3}\tau_t$) should be added when calculating the time taken for charges to reach their equilibrium positions in the base.

Current crowding, or why power transistors have fingers

The analysis of the bipolar transistor has so far assumed that only changes along a line from emitter to collector are worth paying attention to, and that sideways effects do not matter. This is largely true for a small transistor as in Fig. 6.3, perhaps forming part of a memory array. In a high current transistor, passing amps rather than milliamps, transverse effects control the shape of the transistor, leading to designs with many fingers, so that there is a large perimeter to area ratio.

The problem is that in order to obtain a high frequency response and a high gain, the base width is made as small as possible. As a result, the resistance of the base to base current flowing in under the emitter becomes important. The relevant resistance is that between X and Y in Figs. 4.1 and 4.12. Because there is resistance and a current flow, there is a voltage between X and Y, and the forward bias at Y is less than at X. Consequently, the emitter current tends to be kept out of the centre of a large emitter and to be crowded towards the edge. To reduce this current crowding to an acceptable level, the total base current I_B should develop less then kT/e volts between X and Y. To estimate the maximum value of I_B in a design, we shall assume that the base current flows all the way to the centre line of an emitter stripe, instead of being distributed along it. Then

$$R_B = \frac{\rho L}{A} = \frac{1}{eN_B\mu} \cdot \frac{W}{l(2NL)} \leqslant \frac{kT}{eI_B}$$

where l, W and L are defined in Fig. 4.12 and N is the number of emitter fingers. The 2 occurs in the above equation because each

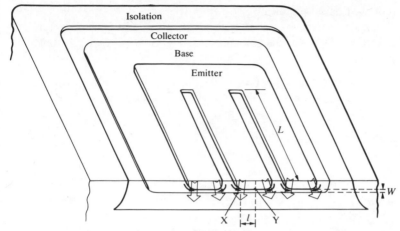

Fig. 4.12. Current crowding in a bipolar transistor. The small arrows show the base current and the broad arrows the emitter current, which tends to flow near the periphery of the emitter.

emitter finger has two sides. The result of current crowding is that high-current transistors have their emitters in the form of many narrow stripes, so that no point on the emitter is far from an edge.

Noise

Noise in colloquial terms is any unwanted signal. Some noise comes from sources outside the circuit (lightning, solar flares, or petrol ignition sparks) and with this we shall not be further concerned. In digital circuits, signals applied on one lead may induce signals on other leads, and a circuit designer has to ensure that the circuit can distinguish reliably between such cross-talk and intended signals. Our attention however, will be concentrated on the irreducible minimum of noise in all circuits which is dependent on first the non-zero temperature of the system, and second the finite magnitude of the charge on an electron.

If a fluctuating current i has a mean value $\bar{i}$, then the fluctuation in i is Δi and $\Delta i = i - \bar{i}$. Now $\overline{\Delta i} = 0$, so $\overline{\Delta i}$ is useless as a measure of the noisiness of i, but $\overline{(\Delta i^2)} \neq 0$, and is known as the variance of i, written var(i). Then

$$\text{var}(i) = \overline{(i - \bar{i})^2} = \overline{i^2} - (\bar{i})^2$$

(Check this by expanding the middle term).

If now a current i has two components i_1 and i_2, which perhaps have been flowing in separate wires and now join, the variance of i is

$$\text{var}(i) = \overline{(\Delta i^2)} = \overline{(\Delta i_1 + \Delta i_2)^2} = \overline{\Delta i_1^2} + \overline{\Delta i_2^2} + \overline{2\Delta i_1 \Delta i_2}.$$

If Δi_1 and Δi_2 are 'uncorrelated', then the third term is zero, so the variances add directly. If the sources of noise are causally independent, then the noise is not correlated. If there is some correlation then $\overline{\Delta i_1 \Delta i_2}$ may be positive or negative. (Exercise: find var(i) when $\Delta i_1 = \Delta i_2$ and when $\Delta i_1 = -\Delta i_2$).

A particular kind of randomness, and hence noise, comes from a train of similar independent events. If the average number of events in a sample is $\bar{n}$, then a result of statistical theory is that noise may be spread over a wide range of frequency, and we use the concept of spectral intensity of the fluctuations in some quantity n in a frequency range $f + \delta f$, which is written $S_n(f)$, where $S_n(f)\, \delta f$ is the average value of the square of the fluctuations in n between f and $f + \delta f$. A further result of statistical theory is that $S_n(f) = 2\,\text{var}(n)$, so that for our random series $S_n(f) = 2\bar{n}$.

A current I of I/e independent electrons per second would have a variance I/e, so the spectral intensity of the fluctuations in the number of electrons per second would be $S_n(f) = 2I/e$, while the spectral intensity of the fluctuations in the current would be

$$S_I(f) = e^2 S_n(f) = 2eI. \tag{4.22}$$

The name shot noise is given to this kind of electrical noise.

The idea of thermal noise applies properly to systems in thermodynamic equilibrium, but may be applied safely to many systems so long as they are not perturbed strongly away from equilibrium. A resistor through which the mean current is zero is a good example of a situation which can be taken to be in thermal equilibrium.

An argument based on classical thermodynamics leads to a set of expressions for the spectral intensity of fluctuations in the current, voltage, or power in a resistance R at temperature T,

$$S_I(f) = 4\kappa T / R,$$
$$S_V(f) = 4\kappa T R, \qquad (4.23)$$
$$S_P(f) = 4\kappa T.$$

The quantized version of (4.23), including the effect of zero-point energy is

$$S_V(f) = 4hfR\{\tfrac{1}{2} + 1/(\exp(hf/\kappa T) - 1)\}. \qquad (4.24)$$

(Check that (4.24) reduces to (4.23) at low frequency, i.e. when $hf \ll \kappa T$.)

The simple versions (4.23) are generally used. They show a flat spectrum for all f. From a practical point of view (4.23) sets a limit below which the noise from a resistor cannot fall, through poor resistors may well have a higher noise output.

Consider now the low-frequency noise properties of a p^+–n junction diode which obeys the equation $I = I_R\{\exp(eV/\kappa T) - 1\}$. When the diode current is I, the hole current injected into the n-side will be $I + I_R$, while the hole current leaking back into the p-side will be $-I_R$. In both of these currents the holes move independently, and contribute full shot noise; for the injected current because the process is random diffusion, and for the leakage because the arrival of holes at the edge of the depletion layer is random, and nothing can arrest their motion once in the high field of the depletion layer. Electron currents are being ignored because the diode is p^+–n not p–n.

The full spectral intensity of current fluctuation on the basis of each current contributing according to eqn (4.22) is

$$S_I(f) = 2eI + 4eI_R. \qquad (4.25)$$

We can compare this with the result derived from the slope resistance of the diode when $I \gg I_R$. The resistance is $\kappa T/eI$, so that from eqn (4.23) $S_I(f)$ is predicted to be $4eI$. The distinction is important, the slope resistance of a p–n junction diode does not act as a thermodynamic noise generator; it generates only half the noise. Extra effects have to be taken into account at high frequency, but for simplicity we analyse only low-frequency effects, both for the junction diode and for the bipolar transistor.

For our model of a p–n–p transistor we start with the Ebers–Moll eqn (4.8). Using the relation $\alpha_i I_{CB0} = \alpha_n I_{EB0}$, the equations may be rewritten for the normal bias where V_{EB} is small and positive and V_{CB} is large and negative,

$$I_E = \{\alpha_n I_{EB0} \exp(eV_{KB}/\kappa T)\} + \{(1 - \alpha_n I_{EB0})\exp(eV_{EB}/\kappa T)\}$$
$$- \{I_{EB0}(1 - \alpha_n)\},$$

$$I_C = \{-\alpha_n I_{EB0} \exp(eV_{EB}/\kappa T)\} - \{I_{CB0}(1 - \alpha_n)\}, \tag{4.26}$$

$$I_B = -(I_E + I_C).$$

There are four distinct terms in braces in eqn (4.26), and they have been indicated by the four groups of holes flowing in Fig. 4.13(a). Each carries full shot noise, and is uncorrelated at low frequencies with the others. The equivalent circuit, Fig. 4.13(b), includes the noise current generators from Fig. 4.13(a), shown in dotted circles, and ordinary circuit elements (the low-frequency hybrid-π model), which operate on signals or noise alike.

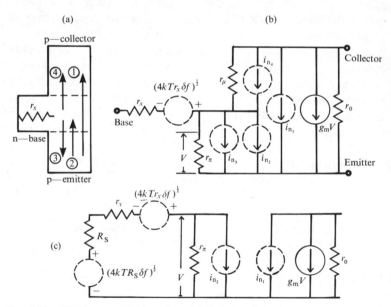

Fig. 4.13. Bipolar transistor noise: (a) hole currents in a p–n–p transistor corresponding to eqn (4.26); (b) the equivalent circuit of (a): (c) a simplified equivalent circuit of the transistor and a signal source.

The values of i_{n3} and i_{n4} are small in Si transistors. When they may be ignored, the value of the other noise currents is

$$i_{n1}^2 = S_{i1}(f)\, df = 2e\alpha_n I_{EB0} \exp(eV_{EB}/\kappa T)\, \delta f = 2eI_C\, \delta f,$$

$$i_{n2}^2 = S_{i2}(f)\, df = 2e(1 - \alpha_n)I_{EB0} \exp(eV_{EB}/\kappa T)\, df = 2eI_B\, df.$$

From the derivation of the hybrid-π parameters, (pp. 107–11) $r_\pi = \kappa T/eI_B$, so that, in agreement with our analysis of a junction diode

$$i_{n2}^2 = (2\kappa T/r_\pi)\, \delta f.$$

A simple circuit suitable for analysis, including a source resistor with its own noise generator, is shown in Fig. 4.13(c). Before we carry out the analysis, we must consider how the noise in a circuit should be assessed.

Noise measures

There have been a number of parameters used to describe noise in a circuit; we shall study the noise figure; other names that have been used are 'operating noise figure', 'operating noise factor', and 'spot noise figure'. The noise figure F for a circuit is the ratio of the total noise power delivered to the output of the circuit when the source resistance is at 290 K to the part of the noise power at the output which is due to the source resistor. The noise figure is thus always greater than one, though in a good circuit it may be only a little greater.

One way to measure the noise figure is indicated in Fig. 4.14. A calibrated noise source, such as a temperature-limited vacuum diode is used. First note the noise power arriving at R_L when the calibrated noise source is switched off. Second, increase the current in the noise source until the power in R_L is doubled. If

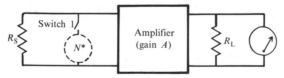

Fig. 4.14.   To measure the noise figure F, the output noise with switch 1 open is noted, and then switch 1 is closed and the calibrated noise source N^* adjusted until the output noise power doubles.

the power in R_L in the first case is O_1, then

$$O_1 = AN_S + AN_A,$$

where N_S is the noise power from R_S and N_A is the noise power from the amplifier, which we describe as if it were all produced at the input. When the noise power arriving at R_L is doubled we have

$$2O_1 = A(N_S + N_A) + AN^*,$$

where N^* is the noise power from the calibrated source. Then $N^* = N_A + N_S$, and from the definition of F

$$F = A(N_A + N_S)/AN_S = N^*/N_S.$$

The value of N^* is known from the calibration of the instrument, and N_S is known from the resistance of R_S and its temperature.

The noise figure for the transistor amplifier of Fig. 4.13(c) can now be calculated, and from the calculation some guidance obtained for the best ways of making a low-noise amplifier.

The noise sources associated with R_S, r_x, and r_π produce a total (noise voltage)^2 of V^2 across r_π, where

$$V^2 = \frac{4\kappa T\, \delta f(R_S + r_x)r_\pi^2}{(R_S + r_x + r_\pi)^2} + \frac{2\kappa T\, \delta f}{r_\pi}\frac{(R_S + r_x)^2 r_\pi^2}{(R_S + r_x + r_\pi)^2}.$$

Across the output resistance r_0, the square of the noise voltage due to all sources is V_{out}^2, where

$$V_{out}^2 = r_0^2(2eI_C\, \delta f + g_m^2 V^2). \tag{4.27}$$

When only the noise from R_S is considered

$$V_{out}^2 = r_0^2 g_m^2 \frac{4\kappa T\, \delta f\, R_S r_\pi^2}{(R_S + r_x + r_\pi)^2}. \tag{4.28}$$

The value of F is the ratio of eqns (4.27) to (4.28), and, in writing it out again, $g_m r_\pi$ has been replaced by β, the small signal current gain.

$$F = 1 + \frac{r_x}{R_S} + \frac{(R_S + r_x)^2}{2R_S r_\pi} + \frac{eI_C(R_S + r_x + r_\pi)^2}{2\beta^2 \kappa T R_S}. \tag{4.29}$$

This result would be the starting point for optimizing the noise properties of a low-frequency common-emitter amplifier. The best values of R_S and I_C for a given transistor and the best value of β can be predicted. It would of course be worthwhile comparing experimental results with theory to show whether the terms which have been ignored in the approximation can be validly omitted.

When passing current all devices have been found to have low-frequency noise power in excess of the value predicted by $4\kappa T\,df$. This is called flicker, or $1/f$ noise. The corner frequency at which the graph of the spectrum bends up from an f^0 dependence to an f^{-1} dependence may be in the range of 100 to 10 000 Hz, and the f^{-1} law appears to hold to the lowest frequencies tested (10^{-5} Hz). The random filling and emptying of a distribution of states at semiconductor surfaces is thought to be the usual cause of this noise, and, fortunately, as manufacturing process control improves for a device, the corner frequency for $1/f$ noise moves steadily down. All that can be done by a user to cope with $1/f$ noise where a small, low-frequency signal is to be amplified is to select low-noise, high-quality devices.

Thyristors

A number of semiconductor devices have been made with four or more layers of p and n alternately. We consider the thyristor as an example. It is made of four regions p–n–p–n, so there are three p–n junctions (Fig. 4.15(a)). It can be switched on by a current pulse to the gate, but once on, can be switched off only

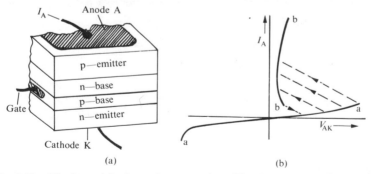

Fig. 4.15. Thyristor: (a) schematic construction: (b) voltage-current characteristic: aa—off, bb—on.

by reducing the anode voltage to or below zero. It is used in low-frequency circuits to control power up to tens of kilowatts by selecting the instant of switching on.

The anode–cathode voltage–current curves for a thyristor are shown in Fig. 4.15(b). Along curve aa the thyristor is off, the anode may be positive or negative relative to the cathode, and any current flow is small. On curve bb, the thyristor is on: the arrows show possible ways a transition between off and on can occur. Notice that at some voltages, the current may be high or low, so we have to find some process inside the device which is different for high and low currents. Several non-linear processes have been suggested, but here we shall examine only one, trap-controlled recombination, though not thereby implying that it is the full explanation in all cases.

As already described on p. 29 there are some impurities which form energy levels near the centre of the forbidden band. Au and Pd in Si are examples. These levels are known as traps, or trapping states, and may 'capture' an electron (Fig. 4.16(a)) which after a delay may escape back to the conduction band (Fig. 4.16(b)) or fall to the valence band (Fig. 4.16(c)). The last process can be also thought of as the further capture of a hole (Fig. 4.16(d)).

Consider a situation where there is an excess electron density n', and the material has T traps per unit volume, of which a fraction f are full. Then the rate of loss of electrons from the conduction band is $n'(1-f)TC$ where C is the rate constant for the capture. If process (b) on Fig. 4.16 is so infrequent that it can be ignored, and process (c) can be described by a half-life τ, then

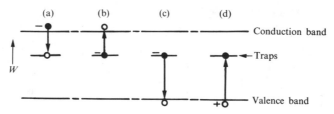

Fig. 4.16. Carrier transitions between traps, and the valence and conduction bands: (a) an electron being trapped: (b) the electron becoming free again: (c) the trapped electron falling to the valence band: (d) the process in (c) may be represented in another way as the trapping of a hole.

for a steady fraction of the traps to stay filled

$$n'(1-f)TC = fT/\tau$$

or

$$f = \frac{n'C}{1/\tau + n'C}.$$

Notice that as n' becomes large $f \to 1$. The recombination rate can now be written

$$\mathrm{d}n'/\mathrm{d}t = n'TC/(1 + n'\tau C).$$

When n' is small, $\mathrm{d}n'/\mathrm{d}t \to n'TC$, so that the rate of recombination is proportional to n'. When n' is large, $\mathrm{d}n'/\mathrm{d}t \to T/\tau$, and hence is independent of n'. The average life of a carrier now becomes longer, and its diffusion distance becomes larger. The effect is observed in bipolar transistors, where the current gain for small emitter currents is low because few traps in the base are filled.

In a thyristor in the off condition, excess carrier densities are low, few traps are occupied, and minority-carrier lifetimes are short. In the on state, excess carriers are high, and the traps are filled so that the minority-carrier lifetimes are much longer, demonstrating that trap-controlled recombination is a process that is non-linear in n'.

One way of regarding a thyristor is as two closely linked transistors in the same block of semiconductor (Fig. 4.17(a)).

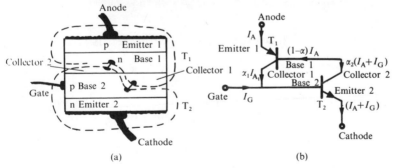

Fig. 4.17. (a) A thyristor can be partitioned into two closely coupled transistors, one n–p–n and the second p–n–p: (b) the currents in the two complementary transistors.

The current between base 1 and collector 2 can be written in two ways, which must be equal (Fig. 4.17(b))

$$(1 - \alpha_1)I_A = \alpha_2(I_A + I_G).$$

Hence

$$I_A = I_G\alpha_2/(1 - \alpha_1 - \alpha_2).$$

If $\alpha_1 + \alpha_2 = 1$, then I_A may have a finite value, even though I_G is zero. The value of α for a normal transistor usually exceeds 0·9, so the required values of α_1 and α_2 in the on condition are low, and occur because the two component transistors run in the saturated mode (see page 115). In the off condition the short life of injected carriers will reduce $(\alpha_1 + \alpha_2)$ to less than unity, so that the anode current is small until switched on by gate current.

Those parts of the thyristor nearest the gate connection turn on first, and the high-density electron–hole plasma necessary to fill the traps spreads by diffusion from the area where it was first formed. As a result there has to be a delay after a gate pulse before a thyristor can pass a high current. A maximum rate of rise of anode current is part of the device specification for a thyristor.

The switch-off time for a thyristor is governed by the need to wait for the plasma to decay to a low enough density for traps once more to be unoccupied. This may take tens of microseconds, and until it has occurred, the voltage across the thyristor must be low, lest the thyristor switch on again. As a result, thyristors are limited to relatively low operating frequencies, being excellent for controlling power at 50 Hz.

The junction-gate field-effect transistor

The junction-gate field-effect transistor, or JUGFET, consists of a conducting channel with drain and source ohmic contacts at the two ends. The channel passes under p–n junction gates (Fig. 4.18(a)) to which a control voltage can be applied relative to the source.

Fig. 4.2(d) (p. 94) shows an experimental set of curves for an n-channel JUGFET relating to the drain current I_D to the drain–source voltage V_{DS}, with the gate–source voltage V_{GS} as a parameter. As the gate is made more negative, less current flows,

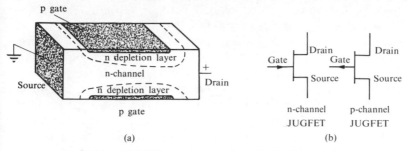

Fig. 4.18.   JUGFET: (a) construction: (b) circuit symbols.

5 V being enough to stop current almost completely. To increase I_D to its maximum value for a given V_{GS}, V_{DS} has to be several volts or more, a noticeably larger value than was required on the collector of a bipolar transistor. Notice also the way the family of curves fan out from the origin rather than breaking from a common envelope as the curves for the bipolar transistor seem to do.

The part of the semiconductor which can carry current is delimited by depletion layers whose boundary can be moved by varying V_{GS} (Fig. 4.19(a), (b)). The device is thus a voltage-controlled resistor. The control voltage is applied across a reverse-biased p–n junction, so very little current flows, and a small power can control much larger powers.

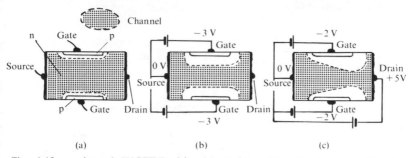

Fig. 4.19.   n-channel JUGFET: (a) without bias: (b) with the gates biased negatively the depletion layers have expanded, reducing the channel width: (c) with negative gate bias and positive drain bias relative to the source, the channel becomes narrower near the drain.

Only two processes are needed to describe a JUGFET: the drift of majority carriers in an electric field, and the variation of the thickness of a depletion layer as the voltage across it is altered. The physics of the device is consequently simpler than the physics of the bipolar transistor, though this is compensated by the need to consider two dimensions in the simplest JUGFET analysis, while a one-dimensional analysis is a useful first stage in describing bipolar transistors.

To allow a reasonably simple analysis, we make two assumptions. A strict theory can show the assumptions are not well fulfilled, but the results fit experiment, and the ideas give a useful picture of the way the JUGFET works.

The two assumptions are 'the abrupt assumption' and 'the gradual assumption'. For the first we assume that the depletion layer ends abruptly at the boundary of the channel. In particular, in the depletion layer there are no free carriers to help carry the drain current, while in the channel the majority-carrier density is equal to the net doping density. Of course in a real JUGFET, the majority-carrier density would fall off smoothly at the edge of the depletion layer, but we shall assume the change is abrupt.

The gradual assumption refers to the rate at which conditions change in passing down the channel from source to drain. We assume that the channel and gate have a length L much greater than the width $2a$ of the JUGFET, so that the changes in the channel width are gradual, and only current flow parallel to the axis of the channel need be considered.

The channel potential varies continuously from source to drain, so the channel-to-gate potential difference V_y also varies, and consequently so does the undepleted channel width (Fig. 4.18(c)). Take as an example an n-channel JUGFET, where the gates are p^+ and the channel is doped with N_d donors. If the width of the depletion layer is x, then at a distance y along the channel

$$x = (2\varepsilon_r\varepsilon_0 V_y/eN_d)^{\frac{1}{2}}.$$

For simplicity, all potentials are internal potentials in the semiconductor. Externally measured potential differences will differ by a constant amount due to contact potentials.

The channel width is $2(a - x)$, and the electric field at a distance y along the channel from the source is $d(V_y)/dy$, so the current is

$$I = -2N_d e\mu_e \frac{d(V_y)}{dy} W(a - x). \tag{4.30}$$

We can integrate eqn (4.30) along the length L of the channel:

$$\int_0^L I \, dy = -2N_d e\mu_e W \int_0^L (a - x) \frac{dV_y}{dy} \, dy,$$

$$I \left|y\right|_0^L = -2N_d e\mu_e Wa \int_{V_{SG}}^{V_{DG}} \left(1 - \frac{x}{a}\right) dV_y. \tag{4.31}$$

If V_0 is the gate-channel voltage difference when $x = a$, then

$$V_0 = eN_d a^2 / 2\varepsilon_r \varepsilon_0$$

and $\hspace{11cm}$ (4.32)

$$x/a = (V_y/V_0)^{\frac{1}{2}}.$$

Carrying out the integral in eqn (4.31) gives

$$I = -\frac{2N_d e\mu_e Wa}{L} \left\{ V_{DG} - V_{SG} + \frac{2}{3}\frac{V_{SG}^{\frac{3}{2}}}{V_0^{\frac{1}{2}}} - \frac{2}{3}\frac{V_{DG}^{\frac{3}{2}}}{V_0^{\frac{1}{2}}} \right\}.$$

This can be written more compactly if we let

$$I_0 = -\frac{2}{3}\frac{N_d e\mu_e Wa V_0}{L},$$

which is the value of I when $V_{SG} = 0$ and $V_{DG} = V_0$, and consequently is the maximum value that I can ever have.

$$\frac{I}{I_0} = 3\left\{\frac{V_{DG}}{V_0} - \frac{V_{SG}}{V_0}\right\} + 2\left\{\left(\frac{V_{SG}}{V_0}\right)^{\frac{3}{2}} - \left(\frac{V_{DG}}{V_0}\right)^{\frac{3}{2}}\right\}. \tag{4.33}$$

Equation (4.33) holds while the depletion layers are narrower than the half-width of the JUGFET. The maximum potential difference between channel and gate will occur at the drain end of the channel, so when $V_{DG} > V_0$ we might expect the depletion layers to overlap and 'pinch-off' the channel. However Fig. 4.2(d) shows that I_D saturates (i.e. stays constant) for large values of V_{DS} rather than falling, so we complete the theory by taking I_D for large V_{DS} to be the value reached when pinch-off occurs, that is when $V_{DG} = V_0$. Then for a pinched-off JUGFET, the characteristic equation is

$$\frac{I}{I_0} = 3\left\{1 - \frac{V_{SG}}{V_0}\right\} + 2\left\{\left(\frac{V_{SG}}{V_0}\right)^{\frac{3}{2}} - 1\right\}. \tag{4.34}$$

As one would expect, when $V_{SG} = 0$, $I = I_0$. A value for V_0 can be estimated from the I_D versus V_{DS} curves. The current should reach a maximum when $(V_{DS} - V_{GS}) = V_0$, and for the example in Fig. 4.2(d), this happens for V_{GS} between $0\,V$ and $-5 \cdot 0\,V$, if we take V_0 to be $5 \cdot 5\,V$.

Real JUGFETs have short gates, so the gradual assumption is not fulfilled, and numerical models show that the abrupt assumption is in error. However the predictions we make about the performances of the device are borne out usefully in practice.

PROBLEMS

4.1. For the Si and Ge transistors whose characteristics are shown in Fig. 4.2 (p. 94) take a quiescent working point at $V_{CE} = 5\,V$ and $I_C = 25\,mA$ and estimate the hybrid-π parameters g_m, r_π, and r_0, and also β.

4.2. Using the hybrid-π model of a bipolar junction transistor with r_x, r_μ and C_μ neglected, find the relation between τ_t and the frequency when the current gain is 3 dB down on its d.c. value for a common emitter amplifier fed from a voltage source.

4.3. If two transistors differ only in that one has a base width which is 90 per cent of the base width of the other, what will be the difference between the base-emitter voltages needed to maintain the same current flowing in the two transistors, under reasonable forward bias?

4.4. Show that in a bipolar transistor where the doping is N in both base and collector, the neutral base width W_B depends on the collector-base voltage V_{CB} as

$$W_B = W_{B0} - \left(\frac{2V_{CB}\varepsilon_0\varepsilon_s}{eN}\right)^{\frac{1}{2}} = W_{B0} - X$$

and that

$$\frac{dI_C}{dV_{CB}} = \frac{1}{4}\frac{X}{W_B} \cdot \frac{I_C}{V_{CB}},$$

where X is the thickness of the depletion layer in the base, and W_{BO} is the total base thickness.

4.5. Find $(S_V(f) \, \delta f)^{\frac{1}{2}}$ and $(S_I(f) \, \delta f)^{\frac{1}{2}}$ for a 1 MΩ resistor and a 1 Ω resistor, taking δf to be 1 MHz and the temperature to be 300 K.

4.6. Plot the way F varies in eqn (4.29) as R_S is varied, and hence find an optimum value of R_S. Take $r_x = 100 \, \Omega$, $r_\pi = 1250 \, \Omega$, $I_C = 1$ mA, $\beta = 50$, $T = 300$ K.

4.7. What properties must a semiconductor possess if it is to be used to make bipolar transistors?

4.8. A silicon p–n–p transistor has base and collector doping concentrations of $1 \cdot 3 \times 10^{23} \, \text{m}^{-3}$ and $1 \cdot 3 \times 10^{24} \, \text{m}^{-3}$, respectively. Its base width is $1 \cdot 0 \, \mu\text{m}$ for $V_{CB} = 0$. Estimate the value of V_{CB} which causes the base width to change by 10 per cent.

4.9. For the device in the previous question, what is the capacity of the base-collector junction if the area is $10^{-8} \, \text{m}^2$. Take $V_{CB} = 0$.

4.10. Find the charge stored in the base of a silicon n–p–n transistor when $I_C = 1$ mA and the base width is 1×10^{-6} m.

4.11. Find the built-in field (average) for the base of a transistor where $n_{BE} = 10^{22} \, \text{m}^{-3}$ and $n_{BC} = 10^{20} \, \text{m}^{-3}$ and $W_B = 3 \times 10^{-6}$ m. What will the transit time be in silicon? What are the minority carriers?

5. Surfaces and interfaces

One kind of interface—that between p- and n-type semiconductor—has already been analysed. Other surfaces also have importance in electronic devices.

In this chapter there is first a treatment of surfaces on the atomic scale, then a discussion of charged layers at an insulator-semiconductor junction, leading to an analysis of the metal-oxide semiconductor (MOS) transistor. Metal–metal junctions introduce metal-semiconductor junctions, used in Schottky-barrier devices. Junctions between different semiconductors are now becoming important, and are briefly discussed.

Surface physics

The surface of a crystal exposed by breaking it in half is clearly not the same as the inside of the crystal. Several methods of analysis confirm this.

The bond model of a crystal describes a surface as having one unjoined bond per atom, or about 10^{19} bonds per square metre (see Fig. 5.1). The bonds can react with neutral particles, atoms or molecules or with charged particles—particularly electrons.

A wavefunction designed to fit an infinite periodic lattice clearly is incorrect outside the crystal. Wavefunctions have been specially devised to fit a periodic potential that stops abruptly. The energy levels of these wavefunctions (Tamm states) differ from the bands of allowed levels which fitted the inside of the crystal. They allow electrons to exist near the surface with energies between the valence and conduction bands. Thus electrons bound on unjoined bonds have a quantum-mechanical description as occupied surface states.

The presence of charges in surface states changes the electrical potential and hence changes the level of the bands of allowed energy, so that the band model of the crystal also reflects the special surface effects. The bands are bent, and much of this chapter is about band-bending. Two points may be noted, (a) half a p–n junction is often worth comparing with a surface layer, and (b) if bands are bent, then $d^2V/dx^2 \neq 0$, so that by Poisson's equation the region is not neutral.

A casually prepared semiconductor surface can have a surface-state density approaching the value of 10^{19} states per square metre quoted earlier, and in addition may have an unknown unstable assortment of chemicals on the surface. Modern semiconductor technology can prepare surfaces which have far fewer surface states, far lower surface charge density in those states, and a stable chemical nature.

The stages in preparing a surface might go as follows: cut the grown crystal with a diamond saw, polish the surface to a mirror finish; etch off a thin layer to remove strained material containing large numbers of structural defects; and finally, after the required doping impurities have been diffused in, deposit a stable insulating layer on the semiconductor surface.

In much of the following discussion we shall ignore the surface states completely, thus simplifying the analysis enormously while

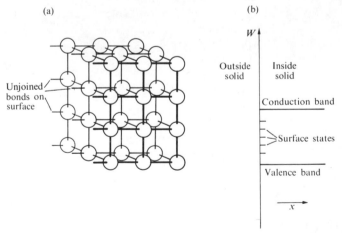

Fig. 5.1. Surface of a crystal: (a) the bond model: (b) the energy-band model.

keeping it accurate enough to be useful. A few devices use a specially prepared active surface, and for these we make a detailed examination of the surface, but in general we concentrate on the regions near the surface, while ignoring effects peculiar to the last few layers of atoms. Surfaces are of interest in many fields—catalysis, friction and detergents are examples—and a fully study of surfaces would draw on contributions from many sciences.

Work function, Fermi level, and electron affinity

The work function ϕ of a crystal surface is the amount of energy required to remove an electron from the middle of the crystal to infinity. There is a further step required to make this definition precise—we must say which electron. For a metal, the choice is easy: we take an electron at the top of the filled levels (Fig. 5.2). At room temperature the levels change from being nearly full to nearly empty in about 100 meV. A typical work function is 4 eV, so for many purposes the Fermi function is a step function, and we can define the work function as the energy required to remove an electron from the Fermi level inside the crystal to infinity, or to give the same idea another name, to the vacuum level.

In a semiconductor there probably are no states at or near the Fermi level, so an electron cannot be removed from them.

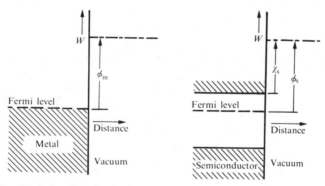

Fig. 5.2. Work functions for both metal and semiconductor are measured from the Fermi level.

Nevertheless, we measure the work function from the Fermi level for a semiconductor, just as we did for a metal. To find how much energy an electron near one of the band edges has to be given to emerge from the crystal we must correct the work function by the difference between the Fermi level and the relevant band edge. It is found in practice that the work function of semiconductors varies as the doping is varied, n-type Si having a lower work function than p-type Si. This variation is nearly all accounted for by the change in the relative positions of the Fermi level and the band edges. The energy required to allow an electron to escape from the bottom of the conduction band is the same for p-type and n-type material. A constant property is useful to work with—we call this property 'electron affinity' χ, (Fig. 5.2).

For a metal
: Electron affinity = work function. Both are measured from the Fermi level.

For a semiconductor
: Electron affinity is measured from the bottom of the conduction band to the vacuum level and is a constant for a given crystal. Work function is measured from the Fermi level to the vacuum level and depends on the doping and on the kind of crystal.

Band-bending and field-induced surface layers

On the positive plate of a charged capacitor there is a positive surface-charge density and on the negative plate an equal negative charge (Fig. 5.3). By considering what states are

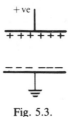

Fig. 5.3.

available for these charges to occupy we can establish how thick the surface layer is which they cause. In a metal there are plenty of states into and from which the extra electrons can be transferred, and the field falls from a high value outside the metal towards zero in a few atomic layers. In a semiconductor there are only a small number of states which, at acceptable energies, can be filled or emptied, and if a large charge has to be accommodated, then the charge on the capacitor extends much further into the crystal.

In Fig. 5.4(b), p-type semiconductor forms the negative part of a capacitor. The usual electrostatic convention that surface charges exist in negligibly thin layers is on this occasion being refined by a more detailed analysis. The charge density at the geometrical surface is zero, so that the electric displacement $D = \varepsilon E$, where $\varepsilon = \varepsilon_r \varepsilon_0$, is continuous in the mathematical sense on entering the surface. As ε is different inside the semiconductor, the electric field inside is not the same as the electric field outside. The displacement falls to zero deep in the semiconductor from its value D outside. Hence the total charge density per unit area, S in the whole surface layer is, by Poisson's equation, equal to D. The bands bend in this case to lower energies near the surface. This removes the valence band from the Fermi level so that there are no holes in the valence band near the surface, though the acceptors stay ionized. The volume-charge density is that of the acceptors, and a useful estimate of the thickness t of this surface layer, if the acceptor concentration is N_A, is

$$teN_A = S = \varepsilon E.$$

This sort of surface layer is called a *depletion* layer, and is similar in many respects to the depletion layer on one side of a p–n junction. For instance, the relation between the voltage V across the depletion layer and the thickness is

$$V = \frac{t^2 e N_A}{2\varepsilon_r \varepsilon_0}.$$

If p-type semiconductor is made the positive plate of a capacitor and hence the electric field is directed out of the surface, the bands are bent to more negative energies near the surface (Fig. 5.4(a)) and there is an *accumulation* of holes near the surface. The charge density increases near the surface as the

p-type semiconductor

(a) *Field is due to negative conductor* Holes are attracted to the surface The field is discontinuous at the surface in all diagrams

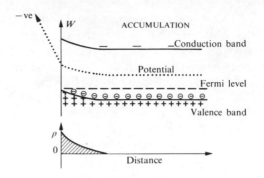

(b) *Field is due to a positive conductor* Holes are repelled from surface, leaving a layer depleted of free carriers

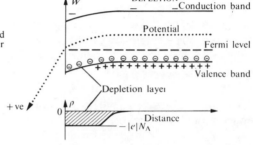

(c) *Strong field due to a very positive conductor* Surface is now n-type. Surface charge is made up of electrons in inversion layer and acceptors in depletion layer

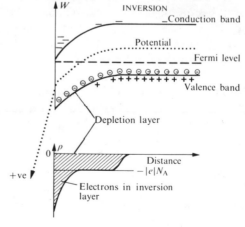

Fig. 5.4. The three types of surface layer for a p-type semiconductor.

valence band approaches the Fermi level, so that the formula describing the thickness of the layer has to include the integral of a charge density which varies with position. The formula is thus more complex than that for the depletion layer.

A strong field into the surface of p-type semiconductor (Fig. 5.4(c)), can cause the conduction band to become the nearer of the two bands to the Fermi level. The surface is then n-type even though the bulk is p-type, and electrons occupy states in the conduction band near the surface in an *inversion* layer. Such a layer forms the conducting channel in many practical devices.

If, instead of p-type semiconductor, n-type is used, corresponding effects occur for the opposite electric field direction. Thus, a field into the surface produces enhancement, a weak field out of the surface produces depletion, and a strong field out of the surface produces inversion. In the inversion layer the majority charge carriers would be holes.

Figure 5.4 needs understanding thoroughly. Check that the sign of the space charge fits with the curvature of the potential, and that charges agree with the gap between Fermi level and band edges. Drawing the corresponding diagrams for n-type semiconductors is a good exercise.

MOSFETs

The idea of using a semiconductor as one plate of a capacitor has been developed into the MOSFET (Metal-Oxide-Semiconductor Field-Effect Transistor) or MOS for short. IGFET (Insulated-Gate FET) is another name, and there are variations if the insulator is not an oxide.

One useful structure is shown in Fig. 5.5. Substantial current only flows between the n^+ drain and source contacts when there is an n-type layer joining them, so that the surface has to be an inversion layer. The gate bias must therefore be positive relative to the source. If the gate bias has not formed an inversion layer, then one of the $p-n^+$ junctions will be reverse-biased, and only a small drain current can flow. The p-type substrate is lightly doped, but is not an insulator and is all at the same potential except in depletion layers, where it abuts onto the three n-type regions. It is biased negatively to both source and drain so that large currents do not flow.

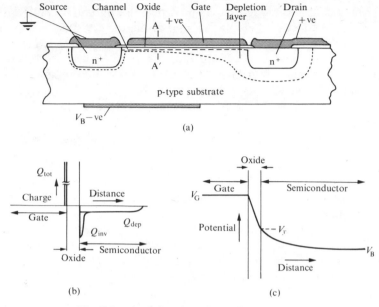

Fig. 5.5.   A section through a MOS transistor.

In the normal set of bias conditions, the depletion layer between source and substrate is thin, the depletion layer between drain and substrate is thick, and that between channel and substrate makes a smooth transition between the two end conditions. The drain must be positive relative to the substrate, and hence is positive relative to the source, or at least only slightly negative.

We will go through an analysis of an MOS transistor twice, once using a simple model, which brings out the main points clearly, and a second time using a more detailed model which includes important features that were omitted in the simple model.

For both models, the mathematics is closely related to that used in describing the JUGFET. The gate and channel are long, and the method is to find the mobile charge, and hence the conductivity of an element of the channel, and then to find the voltage between source and drain by integrating the resistive voltage changes along the channel.

For the simple model, we assume that the potential applied to the gate produces a field in the gate insulator, and that this field all terminates on the mobile inversion charge in the channel. Thus we assume that there are no surface states at the insulator–channel interface, and that there is no change of potential between the channel and the neutral non-inverted semiconductor. Both the assumptions are changed in the more detailed analysis.

In the simple model, the fundamental statement describing the device is

$$Q_{gate} = -Q_m,$$

where Q_{gate} is the charge per unit area on the gate, and Q_m is the charge per unit area in the inversion layer.

Let the channel length between source and drain be L, the oxide thickness be t, the channel width be W, and $V_{(x)}$ be the potential in the channel at a distance x from the source.

Then at x, the field in the oxide is

$$E(x) = V_G - V_C(x)/t.$$

By Gauss's Theorem, the charge at the semiconductor surface layer is equal to the change in the electric displacement D which equals εE. Since the field in the semiconductor is taken to be zero,

$$Q_m = \varepsilon_{ox}\varepsilon_0 E(x) - 0$$

or

$$Q_m = \varepsilon_{ox}\varepsilon_0 (V_G - V_C(x))/t.$$

The channel current I_D is drift current, flowing under the action of the electric field along the channel.

$$I_D = W Q_m \mu \, dV_C(x)/dx$$
$$= \frac{W\varepsilon_{ox}\varepsilon_0\mu}{t}(V_G - V_C(x)) \, dV_C(x)/dx.$$

This equation can be integrated over the length of the channel to give an explicit relation between I_D and the applied voltages which are all referred to the source.

$$\int_0^L I_D \, dx = \int_{V_C=V_S=0}^{V_C=V_D} \frac{\mu W \varepsilon_0 \varepsilon_{ox}}{t}(V_G - V_C(x)) \, dV_C(x)$$

and hence,

$$I_D = \frac{W\mu}{L}\frac{\varepsilon_0\varepsilon_{ox}}{t}\left[V_G V_D - \frac{V_D^2}{2}\right]. \tag{5.1}$$

This equation applies while the field in the oxide can produce some inversion charge at all points in the channel. When V_D is greater than V_G, this is no longer true, so the analysis fails to hold then. The condition $V_D = V_G$ is the point on the I_D versus V_D characteristic where the current has reached its maximum value, and for higher values of V_D the current is found to remain at a constant value, so V_D is replaced in eqn (5.1) by V_G, giving for $V_D > V_G$

$$I_{sat} = \frac{W\mu}{L}\frac{\varepsilon_0\varepsilon_{ox}}{t}\frac{V_G^2}{2}. \tag{5.2}$$

Equations (5.1) and (5.2) show how a given device behaves, which is of value to a circuit builder. They also show a device designer that changes should be made to obtain devices with desired properties.

The values of W and L will be specified by the patterns formed on the surface of a semiconductor slice, as described in Chapter 6. They must be chosen by the designer, and may be termed design parameters. ε_{ox}, t, and μ will be constant for all the devices from the same production line, as the process of ensuring reliable devices is greatly eased if all devices are made in exactly the same way. The name 'process parameters' may be given to these quantities.

A high-current device needs W/L large, while a high-power device needs a large channel area to permit the dissipation of heat, so $W \times L$ must be large in that case.

As a result MOS transistors tend to have gates that are as short as possible, and as wide as is needed for the particular device. Chapter 6 shows some MOS transistor designs.

We can refine the analysis of an MOS transistor by looking more carefully at the distribution of charge near the semiconductor–insulator boundary.

Figure 5.6 shows that there are two other charges to consider besides Q_m. These are the charge Q_{ss} bound on the surface states; and the charge Q_{dep} in the depletion region.

The main effects that are now considered are the depletion

layer between channel and substrate, and the way this is controlled by substrate bias. Other effects are the difference of work function between gate metal and channel substrate, and the band bending needed to invert the channel (Fig. 5.6). Part of the total charge between gate and substrate is mobile inversion charge Q_m; the rest is static charge in the depletion layer Q_{dep} and bound on the surface states Q_{ss}.

$$\varepsilon_{ox} E_{ox} = Q_m + Q_{dep} + Q_{ss}$$

The field in the oxide E_{ox} can be found by adding voltages from point A to point B in Fig. 5.6. We assume that in both source and substrate the difference in potential between Fermi level and majority band is the same (Δv). The potential at B is $V_C(x)$, the potential in the channel at a distance x from the source, and we take the potential at A to be zero.

$$V_C(x) = \Delta v + V_{GS} - E_{ox} t_{ox} + \Delta \phi - 2 \psi_B - \Delta v$$

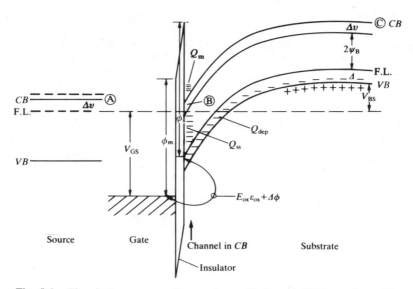

Fig. 5.6. The electron energy diagram for an N-channel MOS transistor. The gate is biased positively and the substrate negatively, relative to the source. The channel is just a little positive relative to the source, so the figure must be showing a part of the channel close to the source.

$2\psi_B$ is the amount of band bending needed to bring the conduction band as close to the Fermi level as the valence band originally was. The voltage between points B and C in Fig. 5.6 controls the value of Q_{dep}.

$$V_{dep} = V_B - V_C = V_C(x) + V_{BS} + 2\psi_B$$

Thus

$$Q_{dep} = \{2eN_A\varepsilon_s(V_C(x) + V_{BS} + 2\psi_B)\}^{\frac{1}{2}}.$$

This now gives an equation for Q_m, using $C_{ox} = \varepsilon_{ox}/t$

$$Q_m = C_{ox}(V_{GS} - V_C(x) - 2\psi_B + \Delta\psi)$$
$$- \{2eN_A\varepsilon_s(V_C(x) + V_{BS} + 2\psi_B)\}^{\frac{1}{2}}.$$

If we take $Q_m = 0$, and $V_C(0) = 0$, we obtain an expression for the threshold voltage V_t of the transistor, when $V_{BS} = 0$.

$$V_t = \Delta\phi + 2\psi_B + (2eN_A\varepsilon_s \cdot 2\psi_B)^{\frac{1}{2}}/C_{ox} + Q_{ss}/C_{ox} \quad (5.3)$$

Notice that by saying $V_C(0) = 0$ at threshold, we are asking that there be no barrier to electron flow from source to channel. Thus the threshold condition is that the conduction bands in source and channel are at exactly the same level. The various components of V_t can be controlled by a manufacturer, and practical values may range from about -2 V to $+2$ V. The special value of V_t needed for CMOS circuits is discussed on p. 173.

The current I_D carried by the mobile charge is $WQ_m\mu\, dV_C(x)/dx$. Integrate this expression to find the total voltage along the channel. For simplicity we take Q_{ss} to be zero from here on.

$$\int_0^L I_D\, dx = W\mu \int Q_m\, dV_C(x)\, dx$$

$$\frac{I_D L}{W\mu} = \int_0^{V_D} C_{ox}(V_{GS} - V_C(x) - 2\psi_B + \Delta\phi)$$

$$+ \{2eN_A\varepsilon_s(V_C(x) - V_{BS} + 2\psi_B)\}^{\frac{1}{2}}\, dV_C(x)$$

$$\frac{I_D}{W\mu} = C_{ox}V_D\left(V_{GS} - \frac{V_D}{2} - 2\psi_B + \Delta\phi\right)$$

$$- \tfrac{2}{3}(2eN_A\varepsilon_s)^{\frac{1}{2}}[(V_D - V_{BS} - 2\psi_B)^{\frac{3}{2}} - (-V_{BS} - 2\psi_B)^{\frac{3}{2}}] \quad (5.4)$$

This expression allows a value to be found for I_D, if the voltages applied to the transistor terminals are known, and it applies as long as Q_m does not fall to zero. Equation (5.4) does not lend itself to an equivalent circuit form, so MOS circuit simulation program packages use a model for the MOS transistor where the drain current is calculated from eqn (5.4) each time it is needed. Notice the use of the term model, rather than equivalent circuit. Equation (5.4) is related to eqn (5.1), having further variables in the first term, and the whole extra second term. The second term becomes less important as N_A decreases, so the simpler equation should be adequate for light doping, if some allowance is made for a threshold voltage.

Real MOS transistors have short channels in which fields are high, so that velocity saturation readily occurs. The mobility in the thin inversion channel is lower than that in bulk material as the particles can make frequent collisions with the two bounding planes of the channel. Both of these are included in circuit simulation packages, the values being calculated each time they are needed.

An exact prediction of MOS behaviour can be made using a computer model which divides the transistors into a large number of smaller regions, and then calculates particle flow and electric field in each region, taking into account that particles flowing out of one region must enter the next, and that electric fields obey similar rules.

MOS transistors are classified into n-channel and p-channel types, depending on the carrier which flows in the channel. A further classification is based on the threshold voltage V_t. If the channel is conducting with $V_{GS} = 0$, it can be further depleted, and the device is a depletion-mode device. If there is no channel when $V_{GS} = 0$, the device is said to be normally OFF, and is called an enhancement-mode device. This kind of MOS transistor needs gate bias of the same sign as the drain bias to permit current to pass. These effects are shown in Fig. 5.7 where the drain current for a large value of V_D is plotted against gate voltage for four kinds of transistor. For example an n-channel enhancement-mode MOS transistor has a positive threshold voltage. Its substrate would be p-type, and the drain current would be of electrons moving towards a positive drain.

n-Channel MOS transistors are faster than p-channel tran-

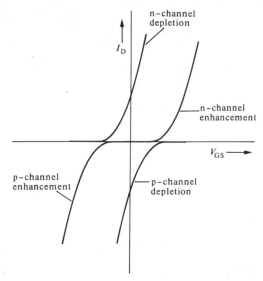

Fig. 5.7. Curves of drain current I_D versus gate voltage V_{GS} for four types of MOS transistor, which have a high constant value of V_{DS} applied. The current carriers in n-channel devices are electrons. A device is an enhancement device if the conductivity must be enhanced above the value for $V_{GS} = 0$.

sistors because the mobility of electrons is greater than that of holes in silicon. In GaAs, the difference in mobility is so great that p-channel GaAs transistors are not considered.

A simple small signal equivalent circuit for an MOS transistor is shown in Fig. 5.8, in the same orientation as Fig. 5.5(a), so as to show which electric process each of the circuit elements represents. The circuit when in use is usually twisted round to look like Fig. 5.9, but the circuit itself has not been modified. The capacitances represent the total capacitance from each terminal to each of the others. The resistance r_{DS} represents the change of I_D, the drain current passing along the channel, when V_{DS} is altered, and the current generator with mutual conductance g_m represents the change of I_D produced by a change of V_{GS}.

As an exercise, the reader could calculate values for r_{DS} and g_m for the MOS transistor in Fig. 4.2(c) for a bias point $V_{DS} = 5$ V, $V_{GS} = 0$ V.

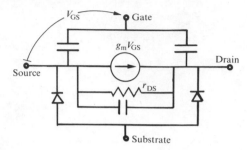

Fig. 5.8. Equivalent circuit for an MOS transistor, drawn with the same orientation as Fig. 5.5. The physical processes can be readily identified.

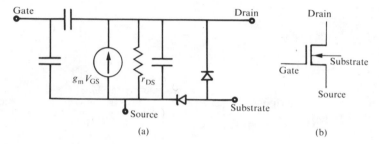

Fig. 5.9. Equivalent circuit for an MOS transistor. This orientation is convenient for circuit analysis.

In this simple equivalent circuit, there are no resistive paths from the gate to the rest of the circuit. This makes the point that the actual resistance is often so high that its admittance can be neglected. In some applications the high-resistance properties are being exploited to the full, and the designer or user might want to know just how high the resistance is. The equivalent circuit could be extended to show this; a value of $10^{13}\ \Omega$ is possible as a gate resistance.

The MOS transistor, as a more recent device than the bipolar junction transistor, has displaced the latter from some applications. Its high input resistance has been mentioned and explains why it is used where very small currents are to be measured. The MOS is used extensively in digital switching circuits. Integrated circuits can readily include MOS devices as their active components, and the large number of logical elements required for a computer may thus be included in a small space. However, for amplifying small signals, linear circuits are easier to build using bipolar transistors, which are thus preferred for some uses.

Charge-coupled devices

Charge-coupled devices (CCDs) use the idea of control by the gate electrode of charge in the channel of an MOS transistor, while discarding the source and drain of the transistor. Figure 5.10 shows an array of gates which are connected to three alternating voltage sources (clocks) ϕ_1, ϕ_2, and ϕ_3, whose waveforms are shown in Fig. 5.10(d). Notice that the voltages are always positive relative to the substrate, which is taken to be at zero potential, and that the waveforms are similar, only being delayed one relative to another.

The positive gate potentials will produce depletion layers in the p-type silicon (as explained on p. 136), but although the potentials are sufficient to produce an inversion layer, there is nowhere for a copious supply of electrons to come from, so the depletion layer stays depleted. If some electrons are deliberately introduced, they will collect under the most positive gate, and this packet of charge carries the signal in a CCD. In Fig. 5.10(b) the voltages on the gates have changed, and the charge has moved part of the way to the new equilibrium position, while in Fig. 5.10(c), the voltages on the gates have changed yet again, and the charge has moved still further.

With the three distinct supplies to the gates, charge can be moved forwards or backwards with a proper choice of wave-forms. Usually only one direction of motion is required, and two clocks connected to two sets of gates are sufficient (Fig. 5.11). The direction of charge movement is then determined by the shape of the gates. Where the oxide is thicker the field is less, and the depletion layer is narrower and of a lesser potential, so the charge packet moves to be under the thin oxide.

Possible ways in which charge may enter the depletion layer are:

(a) by drift from an adjacent part of the depletion layer. This is the normal charge transfer process.

(b) from a forward biased p–n junction. The injection of a charge packet into the depletion layer is done in this way.

(c) by absorption of photons in the depletion layer, producing

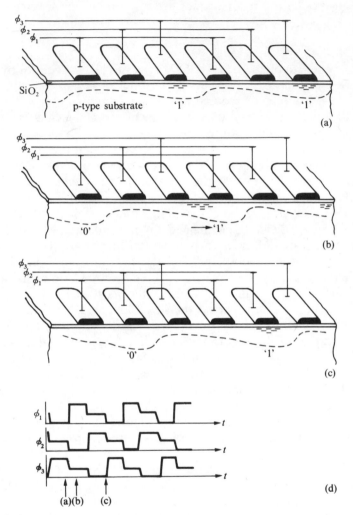

Fig. 5.10. A charge-coupled device: (a) electrons have been drawn under the most positive gates; (b) the voltages on the gates have changed, but the charge packet has had time only to move part of the way under the next gate; (c) the voltages have changed again, and the charges have moved on again; (d) the waveforms for ϕ_1, ϕ_2, and ϕ_3, showing the instants at which (a), (b), and (c) occurred.

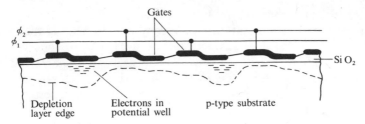

Fig. 5.11. A two-phase CCD. At the instant shown, ϕ_1 is strongly positive (+10 V) while ϕ_2 is low (about 0 V).

electron-hole pairs. In a p-channel substrate, the holes would rapidly leave the depletion layer, and the electrons would remain in the potential well. This is the basis of the CCD television camera.

(d) by thermal ionization in the depletion layer and at the semiconductor/insulator interface.

(e) by diffusion from the undepleted substrate. The depletion layer will collect a current corresponding to that produced by thermal generation in a volume one diffusion length in thickness.

The charge produced in these last two ways is unwanted, and produces noise and distortion of wanted signals.

There are high- and low-frequency limits to the rate at which charge can be moved. The high-frequency limit is set by the time t taken for the last fraction of the charge to move from under one gate to the next. The minimum value of t (i.e. t_{min}) occurs when the electron drift velocity reaches its maximum value v_{max} of about $10^5 \, \text{m s}^{-1}$, so that for a gate of length L,

$$t_{min} = L/v_{max}.$$

A difference of potential of 10 V across $1 \, \mu\text{m}$ can move electrons at v_{max}, but usually the field is lower and the response time is longer.

The low-frequency limit comes from the need to avoid electrons collecting in the depletion layer from the unwanted sources already mentioned. During the time that a charge packet moves from one end of a row of gates to the other, the extra collected charge must not exceed a small fraction of the wanted signal, and so the time for movement along a full row of gates must be less than some limiting value. Silicon with a long minority carrier generation and recombination time is found to be more suitable for making CCDs than GaAs in spite of the high mobility of electrons in GaAs. In practice a charge packet can be moved under a row of 100 gates with a clock rate of 1 to 10 MHz and with a change in the injected charge of less than 1 part in 10^5 for each gate.

After the charge has passed down a row of gates it has to be detected. One way to do this is to collect the charge at a reverse-biased p–n junction, as in a bipolar transistor, and to use ordinary circuit techniques to amplify the current pulse. A charge packet might contain 1 pC arriving over a time of 100 ns, so the expected current is 10 μA. Another way of detecting the charge is to move it to a p-region connected via a metallic track to the gate of a FET, where the collected charge changes the conductance of the FET.

CCDs are used in solid-state television cameras and in binary memories. In the CCD camera, light from the part of a scene corresponding to one line on a TV screen falls on one row of gates. The light produces a pattern of charge in the depletion layer, which is then shifted out, the required video waveform being produced as the charges arrive at the detector element in succession.

In binary memory CCDs a pulse of charge is injected at one end of a row of gates to represent a '1' while the absence of a charge represents a '0'. Once the charge is injected it cannot be sensed until it reaches the end of the row, so the CCD store is a serial memory, not a random access memory (RAM). Because the structure required to store one bit of information is simple, it is possible to build a CCD integrated circuit capable of storing a greater number of bits than can be stored in a MOSRAM. Thus a CCD memory can store data more cheaply but less conveniently than a MOSRAM.

Semiconductor memories

A physical system which can exist in two clearly distinguishable states can be used to store binary data ('0' or '1', 'true' or 'false', 'yes' or 'no'). Read Only Memories (ROMs) have the differences built in during manufacture, Random Access Memories (RAMs) use the two states of a flip-flop or the ability of a capacitor to hold charge and are more properly treated in a textbook on circuits.

In this section we discuss the physical effects that are exploited in those semiconductor memories that can retain stored data after power supplies are disconnected, and where the stored data can be modified.

Figure 5.12 shows one cell of an Electrically Alterable Read Only Memory (EAROM). The cell is a metal-nitride-oxide-semiconductor FET, where the gate insulator has two layers, a thin layer of SiO_2 in contact with the silicon, and a thicker layer of Si_3N_4 on the SiO_2. To store a '1' in the cell, +25 V is applied to the gate for 10 to 20 ms, causing electrons to tunnel from the channel through the thin SiO_2 layer but then to be trapped at the interface between the SiO_2 and the Si_3N_4. The charge at the interface remains trapped when the gate voltage returns to zero, and now induces an equal positive charge in the channel which changes the threshold voltage of the FET.

To erase the stored '1', a large negative voltage is applied to the gate for about 0·1 s. Stored data can be read rapidly from an EAROM by a computer or microprocessor, but a special-purpose instrument is needed to enter or modify data.

Figure 5.13 shows another memory cell, the FAMOS or Floating-gate Avalanche-injection MOS memory. The floating gate is a small area of polycrystalline silicon, totally enclosed in SiO_2. Initially it has no net charge, and it has no effect on the channel beneath it. If one of the p-regions is reverse-biased so that avalanche breakdown takes place, then some electrons are

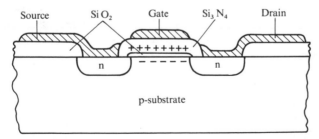

Fig. 5.12.   An electrically alterable read-only memory. The charge trapped at the interface between the SiO_2 and the Si_3N_4 records the stored information.

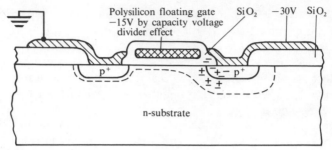

Fig. 5.13. A FAMOS memory cell. Charge can reach the floating polysilicon gate during avalanche breakdown of a p–n junction.

given enough energy to enter the conduction band of the SiO_2, and a few of them reach the floating gate. This negative charge remains on the gate when the avalanche ends, and for years afterwards. As with the EAROM the threshold of the transistor is altered, but the only way to erase the data on the FAMOS memory is to flood it with ultra violet radiation for some minutes, which induces sufficient photo conductivity in the SiO_2 to allow the trapped charge to escape.

This section may have suggested the value of a semiconductor memory cell which could be both read and written into with normal working voltages (<15 V) in less than 1μs, and which would retain the stored information after power supplies were switched off. At present no such device exists. Perhaps a device combining magnetic and semiconductor integrated circuit techniques would do the job.

Metal–metal junctions

The effects occurring when dissimilar metals are brought together are worth studying because they help to explain the metal–semiconductor junction (which is examined in the next section) and also because they are of interest in their own right.

When the surfaces of two metals of different work functions ϕ_1 and ϕ_2 form the plates of a capacitor there will be an electric field between them. If the metals are in contact elsewhere, the Fermi levels will be at the same energy, and the voltage difference in the free space between the two metals will be

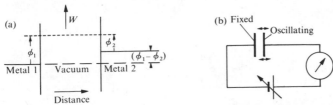

Fig. 5.14. The Kelvin method for measuring work function differences: (a) when the field between the plates is zero, the energy difference between the two Fermi levels is $(\phi_1 - \phi_2)$; (b) the charge on the capacitor is independent of the distance between the plates only when the field is zero.

$(\phi_1 - \phi_2)/e$. This potential difference can have observable effects. It can accelerate electrons, either a deliberately injected electron beam or photoelectrons released from one of the surfaces; and it causes there to be a charge on each capacitor plate. The Kelvin method of measuring work functions (differences of work function in fact) uses this effect (Fig. 5.14). The distance between two plates is varied, and hence the field for a given potential difference. An ammeter in the circuit connecting the two plates will indicate a changing charge on the plates, except when the field is zero, which will occur when an external voltage in the circuit exactly cancels the potential difference $(\phi_1 - \phi_2)/e$. A measurement of the external voltage thus measures $\phi_1 - \phi_2$.

Metal–semiconductor junctions or Schottky barriers

The effects that occur at the interface between a metal and a semiconductor depend on the work functions of the two materials, assuming that surface states do not play a dominant part. To understand this we examine a metal and semiconductor being brought into contact. Figure 5.15(a) shows the two materials well separated. Their Fermi levels are at the same energy; there is a small field in the space between them equal to $(\phi_m - \phi_c)/ed$.

When the metal and semiconductor are brought close enough, the potential difference cannot be taken up in the small space but produces a field in the semiconductor. We may say that the bands are bent (see p. 135). When the materials touch, the whole

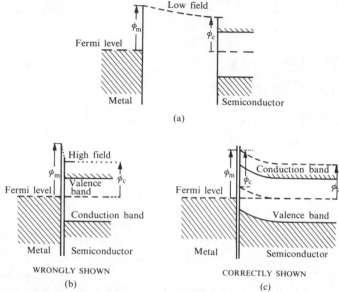

Fig. 5.15. Effects when metal and semiconductor are brought together: (a) when metal and semiconductor are far apart the field between them is low; (b) the sharp change of field shown at the edge of the semiconductor is forbidden by Poisson's equation; (c) the difference of work function is taken up by band-bending in the semiconductor.

of the work function difference $\Delta\phi$ is taken up by the band-bending (Fig. 5.15(c)).

Another way to think about the contact is to say that the electron, in order to leave the metal, has to be given ϕ_m energy, of which ϕ_c is returned on entering the semiconductor. The difference $\Delta\phi$ appears as a change in potential in going from metal to semiconductor. The recipe for drawing diagrams such as Fig. 5.15(c) is therefore as follows.

Start with the metal, draw the Fermi level for the semiconductor in the right place allowing for any external voltage, then draw the energy bands for the bulk semiconductor to fit around the Fermi level. Next locate the bottom of the conduction band at the surface by measuring up by ϕ_m from the metal Fermi level, measuring down by ϕ_c and correcting for the gap between conduction band and Fermi level in the semiconductor. This last stage can be more tidily expressed by saying that, at the surface, the bottom of the conduction band is $\phi_m - \chi_s$ above the metal

Fermi level, where χ_s is the electron affinity of the semiconductor. Finally complete the bent bands in the semiconductor. Their thickness is once again governed by charges in the depletion layer through Poisson's equation. Notice that since the potential in the semiconductor is now no longer uniform, the carrier density will also vary. The local difference between the band edges and the Fermi level must be used to find hole and electron densities.

There are four cases of metal-semiconductor contact worth distinguishing, depending on whether the metal or semiconductor has the larger work function and whether the semiconductor is p-type or n-type. Two of these combinations lead to rectifying junctions and two to non-rectifying or 'ohmic' contacts. We shall examine one ohmic and one rectifying case.

Figure 5.16(a) shows the energy-band diagram for zero bias. The interfacial layer is an enhancement layer, and the holes in it can exchange places readily with electrons in the metal. For small biases of either sign the hole flux will not be balanced and current will flow easily. The contact is known as an ohmic contact, and is one way of making good electrical connections to p-type semiconductor. An ohmic contact to n-type semiconductor can be made when the metal has the lower work function. (Surface states also are important in practical ohmic contacts.)

Case 1. $\phi_m > \phi_c$, *p-type semiconductor*

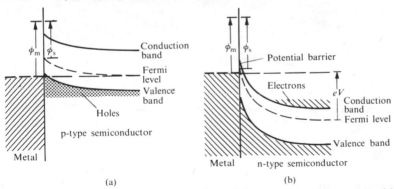

(a)   (b)

Fig. 5.16. Examples of Schottky or metal-semiconductor interfaces: (a) $\phi_m > \phi_s$, p-type semiconductor, no external bias; hole flow is easy, the contact is 'ohmic'; (b) $\phi_m > \phi_s$, n-type semiconductor, diode reverse-biased. The electron flow is impeded by the potential barrier, and the contact is rectifying.

Case 2. $\phi_m > \phi_s$, n-type semiconductor

When the contact is made between n-type semiconductor and a metal of higher work function, a potential barrier is set up between the two layers. The majority carriers in the semiconductor (electrons) now have to flow over a potential barrier of height $(\phi_m - \phi_s)/e$, and their flux is less by a factor $\exp\{(\phi_m - \phi_s)/\kappa T\}$ by comparison with a case where there is no barrier. When an external potential V is applied the barrier is $\Delta\phi + eV$, so the exponential is altered to $\exp(\Delta\phi/\kappa T) . \exp(eV/\kappa T)$. The situation has similarities to a biased p–n junction, but although a $V - I$ relation of the form

$$I = I_0\{\exp(eV/\kappa T) - 1\}$$

seems likely, there are difficulties in establishing a value for I_0.

Metal–semiconductor diodes made in this way are known as Schottky-barrier diodes. Contact between p-type semiconductor and metal of a lower work function also produces a Schottky-barrier diode.

The injection of carriers occurs in the same way with p–n junction diodes and Schottky-barrier diodes, but the recombination of the injected carriers can occur more quickly in the metal than in the semiconductor. The Schottky-barrier diode is consequently very fast, and many microwave detector or mixer diodes are Schottky-barrier diodes. Point-contact diodes, where a metal point is welded to a semiconductor, have long been used for high-frequency rectification, and may have been Schottky-barrier diodes, though without the fact being realized.

Schottky-barrier gate FETs

The Schottky-barrier gate FET or metal–semiconductor FET (MESFET) has a notably simple form of construction, which exploits the possibility of making both rectifying and ohmic contacts on the same semiconductor surface (Fig. 5.17(a)). There are *no* diffusion steps in the sequence of manufacture, so the channel can be very short. Since the mobility of electrons in GaAs is more than four times that of electrons in Si, a GaAs MESFET offers a way of making a device with low electron transit times, and hence a superior high-frequency response.

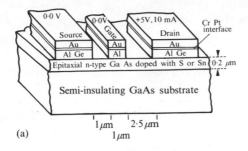

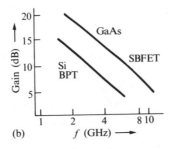

Fig. 5.17.   A Schottky-barrier gate FET: (a) the transistor is completely defined by the metallic pattern of source, gate, and drain; (b) the small dimensions contribute to a superior high-frequency performance of a MESFET amplifier circuit.

The starting material is semi-insulating GaAs, where donors have been compensated (see p. 39) by acceptors. On this is grown an epitaxial layer of high-conductivity n-type GaAs to form the channel. The Ge in the source and drain metallization can act as a donor in GaAs, and helps to make ohmic contacts. The gate is pure Al, and forms a Schottky-barrier, which in use is reverse-biased. A signal on the gate varies the width of the conducting channel, as in a JUGFET, and the equivalent circuit is similar to that of other FETs. The CrPt layers are to prevent the formation of an alloy between Al and Au (see p. 186).

The gain of a MESFET amplifier is plotted against frequency in Fig. 5.17(b), together with the results for a Si bipolar transistor amplifier. The curves show that the MESFET is an effective amplifier at microwave frequencies, and a further reason for its use is that it produces little high-frequency noise.

Semiconductor–semiconductor contacts or heterojunctions

It is possible to make a junction between two semiconductors having different band gaps. A way of ensuring that the regularity of the crystal structure is continued in passing from one material to the other is to grow one substance on top of another crystal. If the two crystals have similar spacing between their atoms, then an 'epitaxial layer' can be grown, and this is the usual method of making a heterojunction.

One currently important example, used in lasers, is a junction between gallium arsenide and gallium arsenide phosphide, $GaAs–GaAs_xP_{1-x}$. In the latter material the As and P atoms play similar parts, and may be present in any ratio—thus varying x. A continuous range of semiconductors can thus be synthesized having steadily changing properties as x goes from 0 to 1.

An important property is the band gap, whose variation with x is shown in Fig. 5.18.

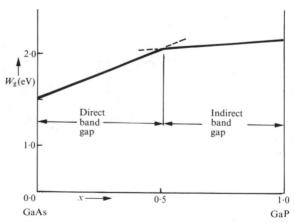

Fig. 5.18. The variation of the band gap W_g of $GaAs_xP_{1-x}$ alloys with the fraction x of As.

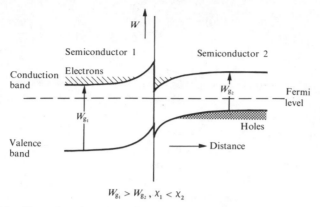

Fig. 5.19. Heterojunction energy-band diagram showing discontinuity in both conduction and valence band edges.

The graph is made of two straight lines. As the composition is varied, every feature in the band structure varies, but at its own rate. The transition from one straight-line segment to another occurs when one local minimum in the conduction band becomes lower than another minimum. A way of describing this is to say that the transition marks the limit between GaAs-like band gap and a GaP-like band gap. As only GaAs has a direct band gap, only samples with $x < 0.5$ are useful for the generation of light.

If there were no noticeable charge bound at the interface, the energy-band picture for heterojunctions could be constructed using the recipe for metal–semiconductor junctions, modified to allow for a step to occur in both valence and conduction bands. (See Fig. 5.19.)

Often one finds that all the data needed to analyse a junction is not available, and the misfit between the two lattices strains the surface, resulting in surface charges, so that the real heterojunction is difficult to analyse in simple terms.

Superlattices (multiple thin-layer materials)

It is possible to grow specimens with hundreds of layers of semiconductor, and to keep the crystal quality high throughout. Where the properties of the semiconductor are altered regularly, a superlattice is said to be formed. Materials made in this way have properties that are not found in normal homogeneous semiconductors, and are expected to lead to devices with improved performance and novel applications.

Three kinds of superlattice are:

Multi-quantum-well superlattices (MQW);

Modulation doped superlattices (MDS);

Strained layer superlattices (SLS).

A multi-quantum-well superlattice has alternate layers with high and low band gap, (Fig. 5.20(a)) so there are multiple heterojunctions. The lattice constants of the two kinds of layer are closely matched to avoid strain at the interfaces between layers. The layers may also be doped to give required properties.

Modulation doped superlattices are based on a single semiconductor, the successive layers being n and p, possibly with intrinsic i layers between them. The question of lattice matching does not arise here, so any semiconductor can be used. The properties of the MDS depend on the thickness of the layers, and of the concentration of the dopants (Fig. 5.20(b)).

Strained layer superlattices (Fig. 5.20(c)) break the rule that the lattice constants of successive layers must be equal. It has been found that thin layers, bounded on *both* sides by material of another lattice constant, will grow without introducing defects into the crystal structure. In a multi-layer structure, alternate layers will be in compression and tension in the plane of the layer, so that the whole structure takes up an average lattice spacing. Because this is different from the natural lattice constant of either of the constituent materials, the properties of both materials are modified from their unstrained values. There is thus access to a range of semiconductor parameters not available in any other way.

The properties of superlattices may be considered in two ways: classically, where the multi-layer structure is just a set of adjacent, distinct layers, and from a quantum mechanical point of view, where the regular variation of properties in the direction normal to the layers introduces a new long-wavelength boundary condition to the wave functions.

A striking classical effect occurs in the MDS structure of Fig. 5.20(b). If extra electron–hole pairs are generated—perhaps by photoabsorption—then the internal periodic electric field causes the electrons and holes to separate into different layers.

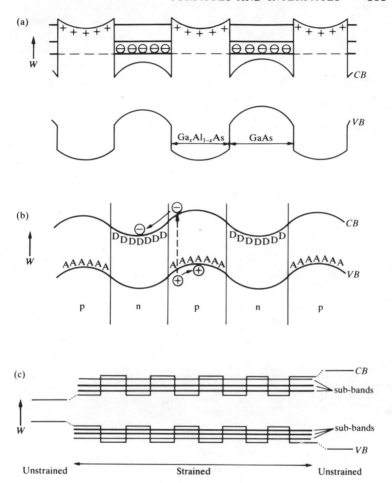

Fig. 5.20. Synthetic semiconductor superlattices. (a) Multi-quantum well superlattice. The free electrons are able to move with high mobility in the undoped narrow band-gap layers. (b) Modulation-doped superlattice. The band gap does not vary. Excess electrons and holes will have difficulty in finding recombination partners. (c) Strained layer superlattice. The superlattice sub-bands are indicated, and will be a stronger effect in narrow layers.

Recombination is then very difficult. Recombination lifetimes as high as 10 min have been measured, in contrast to times between milliseconds and nanoseconds for normal semiconductors.

In an MQW superlattice, if the high band-gap layers are doped and the low band-gap layers are left undoped, then the free carriers from the doped regions will move across to the lower energy parts of the conduction band in the undoped layers. The carriers will then be able to move without suffering collisions with charged dopants. It is thus possible to combine high mobility and high doping, avoiding the fall in mobility shown in Fig. 1.9. The effect is strongest at low temperatures.

Quantum mechanical ideas have been introduced in Chapter 1. Consider first a single-layer quantum well. The density of states in a thin surface has been sketched in Fig. 1.8. When the layer has a finite thickness, the wave functions must fit an integer number of half wavelengths into the width of the well, and this leads to an energy level structure like that of an isolated atom, in the direction normal to the layers. This structure must be combined with the state density from Fig. 1.7 in the plane of the layers, and gives a state density function like that in Fig. 5.21. There are many wells, not just one, the sharp levels are broadened, with each quantum state extending through the whole set of layers.

Thus new energy levels are predicted, and changes in the density of states and consequently of the density of free carriers.

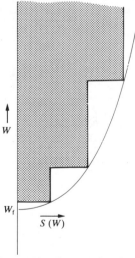

Fig. 5.21.   The density of states for electrons in a layer which is thin, but not of zero thickness. The parabola is the density of states for a normal solid.

These effects are currently being exploited in the active region of semiconductor lasers and other new devices.

PROBLEMS

5.1. The surface-charge density in an inversion layer is $1 \cdot 0 \times 10^{-5} \, C \, m^{-2}$. If the surface mobility of this charge is $0 \cdot 02 \, m^2 \, V^{-1} \, s^{-1}$, what is the surface conductance due to this charge?

5.2. In a Si p-channel MOSFET, the gate potential is $-20 \, V$ relative to the substrate whose doping density is $10^{21} \, m^{-3}$, and a point A on the channel is at $-7 \, V$. If the insulator is $10^{-6} \, m$ thick and has a relative permittivity of $6 \cdot 0$, what is the inversion charge at A?

5.3. In a measurement of work-function difference using the Kelvin method, (p. 153), the plates of the capacitor are of an area $1 \cdot 0 \times 10^{-4} \, m^{-2}$ and are separated by an air gap of $1 \cdot 0 \times 10^{-3} \, m$. If the work-function difference is $2 \cdot 0 \, eV$, what is the current that flows at zero bias when one plate is vibrated at a frequency of $2000 \, Hz$ with an amplitude of $0 \cdot 2 \, mm$.

5.4. One way of arranging a load for an MOS transistor circuit is to use another MOS transistor with its gate connected to the drain. Using load-line constructions analyse the operation of this circuit.

5.5. If there is a bound electron on each of $10^{16} \, m^{-2}$ atoms at an interface, what field change is there at the interface? Take the relative permittivity to be 4 on both sides of the interface.

5.6. In going from SiO_2 to Si the electric field in the SiO_2 is $3 \times 10^3 \, V \, m^{-1}$. What is the field in the Si? Take $\varepsilon_r(Si) = 12$, $\varepsilon_r(SiO_2) = 4$.

5.7. Find the threshold voltage V_t for a p-channel MOS transistor, and the contributions to V_t of the four terms in eqn (5.3). The gate oxide is $1000 \, Å = 0 \cdot 1 \, \mu m$ thick, and has a relative permittivity of 4. The substrate is doped with $10^{22} \, m^{-3}$ donors, and the work functions of metal and semiconductor are 5 and $4 \cdot 4 \, eV$, respectively.

6. Integrated circuits

Introduction

The use of semiconductors in electronics has developed from the cat's whisker detector, using a fragment of a natural mineral, to a computer on a single rectangle of silicon about half a centimetre square. This development represents an engineering achievement which has been made possible by the understanding of the physics of the finished devices and of the manufacturing processes. It was not foreseen, and although further development may be predicted confidently, the physical detail is not known nor are the technical and social effects predictable with any degree of reliability.

Early discussions of integrated circuits suggested that it might be possible to use crystal growth to build a structure whose complete function was that of a required circuit, but where the execution of that function was achieved compactly by a controlled interaction of the parts of the crystal. The present design methods are not based on this concept of integration, and are discussed later in this chapter.

Another false trail was the attempt to build three-dimensional circuits. Present integrated circuits are one device deep, with one layer of metallic connection running above the devices in all but a few cases where further metallic layers are used.

Integrated circuits have been classified by their complexity.

SSI	small-scale integration	1–30 gates
MSI	medium-scale integration	30–300 gates
LSI	large-scale integration	300–10 000 gates
VLSI	very-large-scale integration	10 000 gates upwards
WSI	wafer-scale integration	over a million gates

Some progress has been made with three-dimensional integration. This topic should be watched.

A typical SSI device is a quad, 2-input NAND gate. An 8-bit shift register is an MSI device. A digital watch or pocket calculator employs an LSI device, while large memories or 16-bit microprocessors are VLSI devices. Wafer-scale integration uses an unbroken silicon slice, and selects only the good parts of the circuits for use.

VLSI design methods

VLSI design depends on computers. To see why this is so, consider the number of details that need to be specified in a VLSI device. Take a device with 10^4 gates. Each gate has about ten devices, and each device needs five masks to make it. On each mask, about twenty lines define each device. Combining these numbers, 10^7 lines need to be specified on masks for a VLSI device. This is only a rough estimate, but the true answer must be a number of this size, which may be compared to the 3×10^7 seconds in a year.

Some of the ways that computers are used to help human designers, who are slow and error prone, but imaginative and motivated, are:

(a) By the use of a library of designs for components. A transistor, gate or larger subsystem, once designed, can be copied quickly and reliably many times wherever it is required. RAM and ROM are good examples as each of the cells to hold a single bit can be identical. Further, if a change is made to the design of a subunit, changes to all the places it is used can be quickly made by making the change in the master copy in the library.

(b) By coordinating designers' efforts. If the up-to-date state of a design is stored in a computer, each designer can refer to the others' work as it progresses. VLSI devices take fifty man-years of design effort, spread over perhaps two years, so good communication between designers is a valuable contribution to speedy design.

(c) By simulation of the operation of a design before it is made in silicon. The simulation of logic circuits is effective, and is used. The simulation of devices as represented by an equivalent circuit such as the hybrid-π model for bipolar transistors is also of value. In principle, the function of a set of silicon patterns can be calculated using basic laws such as diffusion and recombination, but for more than a single device, the computation is too large to be possible.

(d) By testing a design for compliance with design rules. This relates to (c) but can go deeper. Each transistor can be checked to confirm that it has power, signal-in, and signal-out connections. The boundary of each pattern can be checked to make sure that basic rules are not broken—for instance, that emitters and bases are not inside an isolation region.

(e) By routing interconnections. It is worthwhile avoiding crossovers where possible, and keeping the length of connections short, so as to reduce capacitance and waste of silicon area. The computer can examine the topology of the system and try many alternative routes to see if there is a simple satisfactory pattern.

(f) By making the masks. The same data-base that feeds the simulation programs and the designers' displays can drive electron beams over photoresist to make the set of masks for the production process. The masks are thus untouched by human hand.

Processing techniques

The manufacture of integrated circuits has exploited the properties of Si and SiO_2. The SiO_2 acts as an insulator to separate metal tracks from the semiconductor; as a dielectric in MOS gates and capacitors; as a barrier to prevent diffusion of doping atoms into selected parts of the silicon; and as a protective skin to keep a contaminating atmosphere away from active devices.

Other semiconductors do not have a tough native oxide. Integrated circuits are not made in Ge because GeO_2 is reactive chemically. The use of GaAs for integrated circuits is being actively studied, and has reached the LSI level. SiO_2, Si_3N_4 or Al_2O_3 are used as insulators for GaAs. The high mobility of electrons in GaAs allows faster gates to be made in GaAs than in

Si, so there is a use for high-speed digital communication devices made from GaAs.

The starting point for integrated-circuit manufacture is a slice of silicon cut from a pure single crystal. The diameter of a slice may be up to 8 inches (20 cm), with costs reducing as the diameter increases. The slices are about 0·2 mm thick and are highly polished. Each slice is eventually divided into individual integrated circuits, which may be between 1 and 5 mm square, so hundreds of silicon chips are obtained from a single slice.

A designer of an integrated circuit first prepares a large-scale set of patterns which defines the features of the circuit. This will be stored in a computer as a set of coordinates for each line of the pattern. When it is decided that the pattern is correct, it is transferred in some automatic way to a physical form. This may involve guiding a light spot over a photographic plate, or steering an electron beam over a sensitive film. The object is the same in each case: the production, in one or more stages of scale reduction, of the same pattern, repeated many times, at the size of the final integrated circuit (Fig. 6.1(a)).

Masks containing many copies of the pattern to the transferred to the silicon may be photographic negatives on glass bases, or may be etched metal films on a glass base. The sequence of operations to make a doped area is shown in Fig. 6.2. The silicon is prepared for the operation by first having a layer of SiO_2 grown on its surface by heating in air, possibly in the presence of water vapour. This layer is likely to be about 1 μm thick.

Then the SiO_2 is covered with a layer of varnish. A drop of the varnish (photoresist) is put in the middle of the silicon slice, which is then spun to spread the photoresist smoothly over the surface. The useful properties of photoresist, SiO_2, and Si are:

Photoresist is hardened by UV;
Hardened photoresist does not dissolve in the right paint stripper, while the unhardened photoresist does so readily;
Hardened photoresist does not dissolve in HF(hydrofluoric acid);
SiO_2 does dissolve in HF;
Si dissolves only slowly in HF.

Figure 6.2(a) shows a mask in place over a slice of prepared Si. The slice of Si has been illuminated with UV through the mask

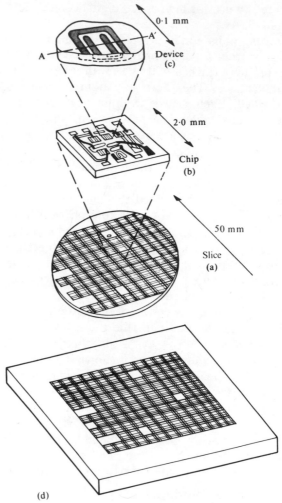

Fig. 6.1. Integrated circuits: (a) a slice of single-crystal Si; (b) one integrated-circuit chip; (c) the pattern on the top of one transistor forming part of the integrated circuit in (b); (d) the mask used for the metallization.

and the photoresist has been hardened except where it was below an opaque region of the mask. Figure 6.2(b) shows the unhardened photoresist dissolved away, exposing an area of SiO_2. Figure 6.2(c) shows the slice after the exposed SiO_2 has been etched away by the strong acid HF. In Fig. 6.2(d) the

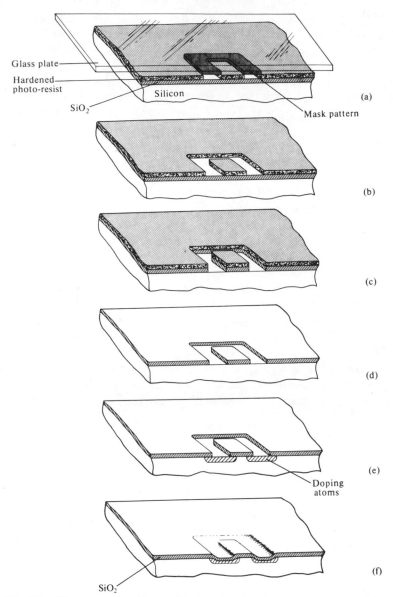

Glass plate

Hardened photo-resist

Silicon

SiO$_2$

Mask pattern

(a)

(b)

(c)

(d)

Doping atoms

(e)

SiO$_2$

(f)

Fig. 6.2. The stages of making a 'diffusion': (a) exposure of the photo-resist; (b) the photo-resist patterned; (c) the SiO$_2$ etched away (d) the rest of the photo-resist removed; (e) doping atoms diffused into the silicon; (f) another layer of SiO$_2$ grown on the surface.

photoresist has done its job, and has been all removed, leaving the SiO_2 to act as a selective barrier to doping atoms reaching the Si surface and diffusing in it.

Figure 6.2(e) shows a doped region. The impurity atoms have diffused both downwards and sideways and the sideways diffusion must be allowed for in the mask design. Diffusion depths would not exceed $10\,\mu m$, and 0·5 to $2\,\mu m$ would be more common. Figure 6.2(f) shows a new oxide layer grown over the whole surface, so the slice is ready for the next stage of processing. The oxidation consumes some of the Si and results in an uneven surface, with the SiO_2 being thicker in some places. The varying thickness of SiO_2 can be seen in magnified images of integrated circuits, as it gives a range of interference colours. The history of processing can thus be seen on the surface of an integrated circuit.

The sequence of steps in Fig. 6.2 is known as a diffusion, and may be repeated several times in making an integrated circuit. When the object is to make an SiO_2 pattern (for instance for use as a gate insulator), the final diffusion step is omitted. To make a pattern of metallic tracks on the surface of an integrated circuit, metal is evaporated over the whole surface and then covered with photoresist, which is then used to protect the metal rather than SiO_2. The unwanted metal is then etched away.

Several masks are needed to make each integrated circuit. The set needed to make the bipolar transistor shown in Fig. 6.3 are

Isolation   (deep diffusion)
Base   (shallow diffusion)
Emitter   (shallow diffusion)
Window   (aperture in oxide)
Metallization (track separation)

The disc of p-type silicon has an epitaxial layer of n-type silicon grown on it. Using the method shown in Fig. 6.2, deep p-type diffusions are formed round each region where a transistor or any other component is to be made. The transistor is thus surrounded by p-type semiconductor on all sides. These diffusions will spread at least a distance equal to the thickness of the epitaxial layer to each side and so require a large area. The next diffusion is the base diffusion, where acceptors are diffused only part of the way through the epitaxial layer. The depth of diffusion is controlled by temperature and time. The base diffusion will be cooler and/or quicker than the isolation diffusion. Notice there is no collector diffusion, as the collector is formed from the starting material.

The next diffusion to make n^+ regions is used for two purposes. The main one is to form the emitter of the transistor by converting part of the p-type base back to n^+ material which is suitable for an emitter. The area of the emitter diffusion is the useful working area of the transistor, and is much less than the area that has to be allowed where the need for contacts and isolation is taken into account; as can be seen in Fig. 6.3. The base width has been shown in Chapter 4 to control the speed of response of the transistor, and the base width is determined by the difference in the depth of the emitter and base diffusions.

The secondary use of the emitter diffusion is to add an n^+ layer on those parts of the n-type collector where there are to be metallic contacts. The n^+ metal combination forms a lower resistance contact than n–metal, particularly if the metal is Al which can act as an acceptor in the surface of the semiconductor.

Windows in the oxide layer must be opened to permit contact from a metallization layer, and these are defined by the window mask. The final mask is the metallization mask which defines the pattern of metallic interconnection between components, and from the circuit to the bond pads.

MOS circuits differ from bipolar transistors in the processing steps in their manufacture. Figure 6.4 shows one way of making an n-MOS inverter. An n-channel enhancement-mode transistor forms the active device and the load is another n-channel MOS transistor which has its gate tied to the source, so that V_{GS} is fixed at the value zero. The load transistor has been made to have a negative threshold voltage by an acceptor-ion implant so it is a depletion-mode device. It is longer than the active transistor in order to obtain a suitable value for its resistance.

The sequence of manufacture starts with the ion implant and then the growth of the thin oxide above the future channels. Next the n^+ diffusions for the source and drain of each transistor are performed. Then more silicon is deposited over the whole slice with those regions needed to make gates and contacts to source and drain being preserved while the rest is etched away. This silicon is polycrystalline, is known as polysilicon, and is not

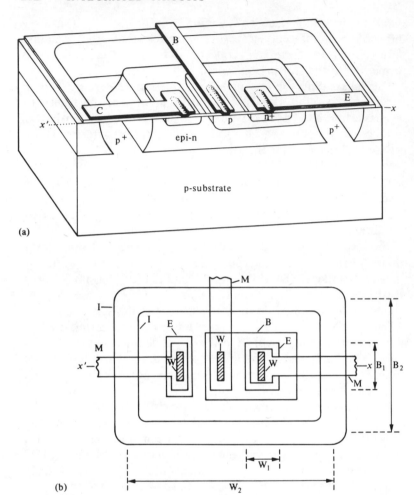

Fig. 6.3. A bipolar transistor made using the silicon planar process: (a) view of half the completed transistor; (b) the patterns for the set of masks. I, isolation mask; B, base diffusion mask; E, emitter diffusion mask; W, window mask; M, metal interconnection mask. The current-carrying area is $B_1 \times W_1$ which is much smaller than $B_2 \times W_2$.

suitable for the base or channel of transistors. The silicon is serving here as a conductor in place of another metal. The polysilicon is then covered with a deposited (not grown) layer of SiO_2 through which windows are opened for the upper layer of metal used for interconnections. Where transistor action is not

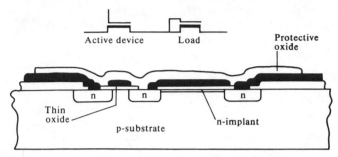

Fig. 6.4. An n-channel MOS amplifier. Ion implantation is used to change the threshold voltage of the load transistor so it is conducting when $V_{GS} = 0$.

required the oxide is much thicker so that unwanted channels are not formed.

Figure 6.5 shows a CMOS inverter. Here the distinction between active device and load has been lost, as both devices are active, each acting as a load for the other. The n-channel device is similar to the active device in Fig. 6.4 and the p-channel device is its complement. One of the two devices can be made in the substrate material, while the other must be formed in a region of opposite doping—in this case a p-well in the n-substrate. Both devices must have threshold voltage lying between V_{SS} and V_{DD}, so they must both be enhancement-mode devices.

Figure 6.6(a) shows the load-line construction for an n-MOS amplifier with a resistor as load, and with either of two n-MOS transistors as load. If a long narrow n-MOS depletion-mode transistor with gate connected to source acts as the load, a large output voltage swing is obtained, with the ON current being acceptably small. Figure 6.6(b) shows the operation of a CMOS

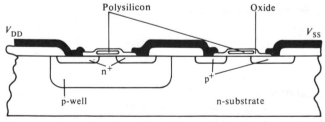

Fig. 6.5. An n-channel and a p-channel transistor makes a CMOS inverter. Each has a polysilicon gate.

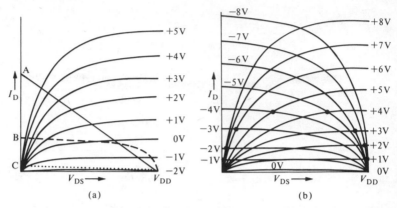

Fig. 6.6.   Loads for CMOS circuits: (a) n-MOS characteristics showing a resistor (A), an n-MOS depletion-mode transistor of the same shape as the active device with its gate connected to its source (B), a longer transistor than (B) with a higher resistance (C); (b) CMOS curves. The heavy dots mark the load line of the full circuit. Note that for high and low voltage output the current is low.

circuit. The dotted load line is the locus of points defined by the conditions that the two gates are at the same potential, and the sum of the source–drain voltages equals the supply voltage. Both OFF and ON states have an effectively zero current.

The layout and logic diagram of a CMOS NAND/AND gate are shown in Fig. 6.7. The regularity of the diffusion patterns may be noted, and also the way the metallic connections run across gates and diffusions to avoid the need for cross-overs.

In addition to MOS and bipolar transistors, other components can be made in integrated circuits. Diodes are straightforward, as transistors can be converted readily to diodes. Probably the best way is to connect base and collector together for one terminal of the diode and to use the emitter as the other terminal. Resistors can be made in any of the layers of semiconductor, and usually the length is many times the width (Fig. 6.8). Capacitors can be formed using the depletion-layer capacitance of a p–n junction, or as in Fig. 6.8 using the thin oxide of MOS gates. Only low values of resistance and capacitance can economically be made, as the large values (>1 kΩ or 10 pF) take up more space than many transistors combined. There is pressure on designers to find circuits that use transistors and diodes only, avoiding resistors and capacitors where possible, and inductors completely.

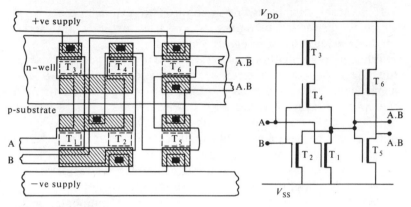

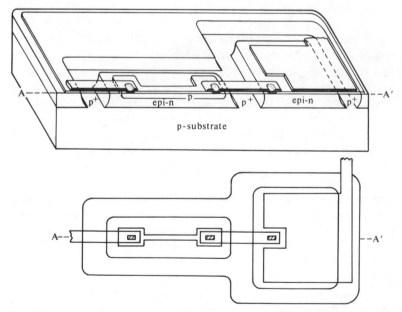

Contents from metal to source or drain
Gate regions
n– or p– diffusions

Fig. 6.7.   CMOS layout and logic diagram for a gate with NAND/AND outputs. Notice that each signal goes to the gate of both an n-channel and a p-channel transistor. Cross-overs are present in the logic diagram, but have been avoided in the layout.

Fig. 6.8.   A resistor and capacitor made using the silicon planar process. The masks are the same sequence as those in Fig. 6.3.

It is often possible to find a solution to an engineering need which avoids the use of inductors, and uses methods that can be readily executed using integrated circuits.

When a component is made, its properties will not be those of the ideal equivalent circuit. The resistor in Fig. 6.8 has capacitance between the base diffusion forming the resistor and the epitaxial layer. The capacitor in Fig. 6.8 has series resistance and also capacitance to the substrate. These unwanted components can be designed to be small, and included in simulation of the behaviour of the circuit. It is also possible to build unwanted transistors, as for example the p–n–p transistor in Fig. 6.3 which is made from the p-substrate, n-epilayer, and p-base. If such a transistor exists, it must never have any combination of voltages applied which would let it pass current.

Isolation methods

The designer of an integrated circuit needs to be able to calculate the performance of his proposed design before it is built. The method used to do this is to construct the circuit from components which interact with each other as little as possible, and to use equivalent circuits to calculate the performance with parasitic elements being added to the equivalent circuit to describe the residual interactions. As far as possible, then, an integrated circuit is not integrated but an assembly of isolated circuit elements in the same silicon chip.

The main method of isolation is junction isolation. An example of this is shown in Fig. 6.3 for a bipolar transistor. The n-type collector is completely surrounded by p-type substrate. The p–n junction between collector and substrate is kept reverse-biased (or at worst at zero bias). A suitable way of doing this is to connect the substrate to the most negative supply to the circuit. Resistors and diodes can be isolated similarly.

Where two devices have their outermost layer at the same potential they need not be isolated from each other. An example is the two transistors which form a Darlington pair, where the two collectors are at the same potential.

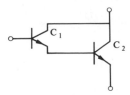

A second, less common, method of isolation exploits the epitaxial growth of silicon on synthetic sapphire, which is Al_2O_3. The isolation is excellent, as the transistors can be separated by any required distance. Figure 6.9 shows two transistors, a p-channel enhancement-mode device, where, at zero gate-voltage, this channel is depleted of free carriers by threshold-voltage effects, and a normal n-channel device.

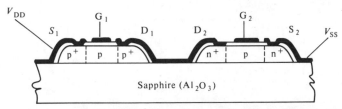

Fig. 6.9.   Silicon-on-sapphire. The two transistors need no further isolation apart from the removal of the silicon between them. Two kinds of MOS transistor can be formed by doping of the ends of the silicon islands p^+ or n^+.

Limits to miniaturization

The number of devices in an integrated circuit has been rising steadily ever since more than one device was included in a monolithic circuit. Readers will have a chance to see whether this growth in circuit size continues, or whether the graph in Fig. 6.10 shows signs of reaching some horizontal asymptote.

The size of a chip which can be made free of any defect in the perfection of the silicon has been increasing steadily, and logical devices are now made about $1\,cm \times 1\,cm$ where necessary. (In contrast, thyristors are made with one thyristor taking the whole of a 2- or 3-inch diameter slice, but in that case, a defect means only that a small part of the surface carries no current, and does

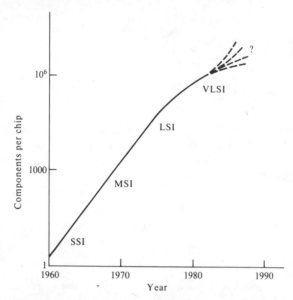

Fig. 6.10. The plot of the number of components per chip against time (the Moore curve) has doubled each year for about twenty years. There are now signs of a lower rate of increase.

not cause failure of the complete device.) The main contribution to miniaturization has been reduction in the size of individual devices, associated with the reduction in the size of features in the masks. In this section the effect on the properties of a circuit of reducing its size by a scale factor are considered, followed by a discussion of physical effects that seem to set a limit to continued miniaturization. These limits are perhaps best thought of as challenges to ingenuity.

For MOS devices, consider the effect of making all linear dimensions in a design smaller by a factor $1/K$ (K is >1). A decision has to be made how to scale electrical quantities. One choice is to scale voltage so that fields are held constant, and breakdown is approached to an equal extent in circuits before and after scaling. The doping is scaled by a factor K. From this point, many results follow using simple physical laws such as $V = IR$ and Power $= VI$ as shown in Table 6.1.

TABLE 6.1

Scaling of MOS circuits

Property	Factor	Comment
Device dimension-L, W, t_{ox}	$1/K$	
Doping concentration	K	Assumed
Voltage	$1/K$	
Current $= \dfrac{V}{R} = \dfrac{VA}{PL}$	$1/K$	
Resistivity	$1/K$	
Resistance	1	
Logic gate area	$1/K^2$	
Depletion and oxide capacitance $= A/t$	$1/K$	
Energy in capacitance (E)	$1/K^3$	
Delay time $= \tau = R_{on}C$	$1/K$	Good
d.c. power/gate (VI)	$1/K^2$	Good
Frequency-dependent power per gate E/τ	$1/K^2$	Good
Frequency-dependent power per chip	1	OK
(Logic operation/s) on a chip $1/(\tau A)$	$1/K$	Good
(Logic operation/s) on a fixed-size chip	$1/K^3$	Very good
Output power from device	$1/K^2$	Bad

For metallic interconnections, similarly scaled for area

Resistivity of metal	1	Fixed
Line resistance	K	
Line-voltage drop/V_{supply}	K	Bad
Line response time	1	
Line-current density	K	Bad

The calculations set out in Table 6.1 show that there is a considerable advantage to be obtained in making MOS devices smaller, with problems in the interface between the miniature world on the chip and the external circuits, and also in the properties of the metallic connectors on the surface of the chip.

A similar analysis can be made for bipolar circuits, though voltages are already at the minimum value in some circuits.

The factors which work against miniaturization may be divided into two groups; those that affect the manufacture of a circuit and those that affect the performance of a circuit that has been successfully made. At present, it is mostly effects in the first group which are of practical importance.

The finest detail on a mask that is prepared by photographic methods is set by diffraction at a size equal to one wavelength of the radiation that is used to make the mask. This sets a limit to the feature size on a mask of about 1 μm. Allowance has to be made for some error in aligning successive patterns, with the result that the minimum feature size on a chip is several times the size of a small feature on a mask.

Small features can be made by using ultraviolet radiation without a major change of technique, or by using electron-beams or X-rays where the wavelengths are sufficiently short that diffraction no longer sets the limit to resolution.

A problem associated with mask resolution is the bowing of silicon slices. Silicon has a higher coefficient of expansion than SiO_2. When an oxidized slice is cooled after oxidation, the silicon shrinks more than the SiO_2, so the slice is curved. An extra allowance has to be made in mask-alignment tolerances for this effect.

The voltages in a circuit impose restrictions on its design. Two fixed values are $\kappa T/e = 25$ mV at 300 K, and the band gap for silicon = 1·1 V. Although both can be changed (by drastic cooling and the use of other semiconductors) the convenience of using silicon at room temperature is so great that an alternative will only be used if its advantages are overwhelming.

A bistable circuit may be thought of as a two-state system with an energy barrier W separating the two states. For transitions driven by noise to be few, the voltage corresponding to the barrier needs to be about $50\,\kappa T/e$ or 1·2 eV. Consequently the minimum power-supply voltage needs to be about this value. The argument that leads to this conclusion may be illustrated from the following example.

Assume that a gate has a switching time of T_S, and that it is required to make no more than one spontaneous transition in time T. The effect of the energy barrier is to introduce a factor $\exp(-W/\kappa T)$ in the relation between the number of chances of a transition and the number of successful transitions

$$\frac{1}{T} = \frac{1}{T_S} \exp(-W/\kappa T).$$

If $T_S = 10^{-9}$ s and $T = 3 \times 10^7$ s = 1 year, then W is about 1·0 V. Since a circuit will have 10^3 to 10^6 gates, an extra safety factor should be introduced, leading to minimum supply voltages between 1 and 2 V. These voltage levels are already in use in current-injection logic (I^2L) circuits, though MOS circuits as used in calculators and watches usually use at least 3 V. It can be seen that the scope for further miniaturization is limited here.

The phenomenon of breakdown also sets a limit, because breakdown occurs at lower voltages as the doping density is increased. Avalanche breakdown is not a problem at low voltages, as the total voltage is too small for any electrons ever to acquire enough energy to produce an electron–hole pair. Zener breakdown by tunnelling does occur and links the maximum doping density to the maximum voltage that can be used. The product of the voltage and doping density will be less than 5×10^{24} V m^{-3} for Si. Concentrations of ten to one hundred times this value can be reached for the common donors and acceptors, but in effect the range of doping above about 10^{24} is not available to device designers.

Metallic conductors in integrated circuits must not be asked to carry more than a certain current density, so they may not always be made as small as the designer would wish. Electro-migration is the movement of atoms in a current-carrying conductor in the direction of motion of the charges (usually electrons but sometimes holes). At high current-density, this can result in the breaking of the conductor as the metal is swept away from a constriction. In Al, the effect is significant at 10^{10} A m^{-2}, so the figure of 10^9 A m^{-2} might be taken as a safe maximum current density. For a current of 10 mA, the required cross-section of Al is 10^{-11} m^2, which could be 1 μm $\times$ 10 μm.

The small number of impurity atoms in a small semiconductor device introduces statistical uncertainties into the properties of the device. The doping atoms have diffused independently, so the number in equivalent volumes is described by a Poisson distribution. Hence if the mean number of atoms in a volume is N, the standard deviation is $N^{\frac{1}{2}}$. The fractional uncertainty is then $N^{\frac{1}{2}}/N = N^{-\frac{1}{2}}$.

Example. The required uncertainty in a region of a semiconductor device is 10 per cent. How many doping atoms must be in the device?

$$N^{-\frac{1}{2}} = 0 \cdot 1, \text{ hence } N = 100.$$

There must be 100 doping atoms in the region for there to be reasonable confidence in the number being within 10 per cent of its nominal value. If $N = 10^{23} \text{ m}^{-3}$, the volume could be a cube of side 10^{-7} m.

This can be compared to the base of a transistor which may be no more than 10^{-7} m in width, though having a greater area. In practice, doping by diffusion or crystal growth only produces uniform doping over a range 1–1·5, though doping by neutron irradiation can produce uniformity to about 1–1·02.

Soft errors

A soft error is a temporary fault in a logic circuit, as opposed to a hard error which is permanent and can be detected every time a test is made. One cause of soft errors is the production in silicon of electron–hole pairs by single high-energy particles— cosmic rays or radioactive decay products.

Radioactive decay of U and Th (impurities in integrated circuit packaging), gives 5 MeV α-particles. Each α-particle produces electron–hole pairs at a rate of 1 pair for each 3·6 eV, so that about $1·5 \times 10^6$ pairs are produced over a track length of 30 μm.

Semiconductor RAM and CCD memories store a logic one as a charge on a capacitor. If we assume that occasionally one-fifth of the charge is collected because the direction of motion of the α-particle keeps it in or near the depletion layer then we can calculate the size of a depletion-layer capacitor which would store 3×10^5 electrons.

We know

$$Q = CV = \varepsilon AV/d;$$

hence,

$$A = \frac{Q}{\varepsilon} \cdot \frac{d}{V} = \frac{eN \cdot d}{\varepsilon V}.$$

If we take

$$N = 3 \times 10^5 \text{ electrons, } d = 10^{-6} \text{ m},$$

$$\varepsilon = 10^{-10} \text{ F m}^{-1}, \text{ and } V = 5 \text{ V},$$

then

$$A = 3 \times 10^{-11} \text{ m}^2.$$

Thus an area $3\,\mu\text{m} \times 10\,\mu\text{m}$ stores 3×10^5 electrons. An α-particle can discharge this capacitance thus producing a false logic output. It has been found that as the stored charge in practical memory cells has been reduced from $2 \cdot 5 \times 10^6$ to $0 \cdot 5 \times 10^6$ electrons the soft error rate has risen from 10^{-8} errors per cell to 10^{-3} errors per cell per hour.

Reliability

The reliability of an electronic system is a measure of the extent to which it performs its intended function. Many devices, circuits, or systems are reliable for a period, and then fail, when they are said to have come to the end of their 'life'. A user will wish to know the expected life of a component he is purchasing, so manufacturers must ascertain the expected life, and know what must be done to produce devices with long enough lives. A related problem is that of ensuring that a device performs its intended function, i.e. 'meets its specification' when first used.

In practice, most semiconductor devices never fail; they are discarded first, so their actual life is not known. However, the fraction of the devices failing in some long interval can be quoted, and values in the range $0 \cdot 01 - 0 \cdot 0001$ per cent per 1000 hrs might be regarded as typical. To establish such figures directly takes many devices and much time, though it is, on occasion, undertaken. By testing devices more strictly in a more severe environment, failure can be hastened, and some evidence garnered rapidly for the reliability of the device. However, relating the evidence from such an accelerated life-test to the equivalent figures under normal conditions is a difficult problem. An experimental solution is to vary the severity of the extra stress—commonly temperature—and extrapolate back to standard conditions. If, as often happens, failure occurs by some route in which an energy barrier ΔW (preferably identified) has to be overcome, then the rate should vary as $\exp(-\Delta W/\kappa T)$. If it does so, then there is some reason for believing the extrapolation to normal conditions.

For many devices and systems the bathtub curve (Fig. 6.11) summarizes the failure rate as a function of time. There is a relatively high 'infant mortality', a lower steady failure rate, and possibly an increase in the failure rate as items become

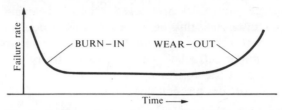

Fig. 6.11. The 'bathtub curve'. For many systems, the failure rate is observed to fall to a steady value after an initial period. Later the failure rate rises as the system becomes 'worn-out'.

'worn-out'. The early failures can sometimes be weeded out by a burn-in test, where all devices are run at full power for a short time and then tested.

In general it may be said that perfect reliability is unattainable, but that any desired standard may be achieved by taking pains enough. When a device is newly introduced and ill understood, its manufacturing yield is low, cost is high, and reliability low. As the manufacturing techniques come under better control, the physics of the device becomes better known, the yield increases, and the balance between cost and reliability is brought ever lower, though the best devices still cost most.

Failure modes

The reason for the failure of a semiconductor device is unlikely to be found in the bulk semiconductor itself. Changes in the properties of the semiconductor can occur, but some such cause as bombardment by high-energy radiation is required. (The radiation can reduce minority-carrier lifetime by introducing defects in the lattice which act as traps.) The p–n junction which is the heart of many semiconductor devices is also very stable, as the distinction between p- and n-type semiconductors is, speaking chemically, very slight. In evidence of this, the temperature at which dopants diffuse is much higher than the usable temperature of devices. (An exception is the very small impurity Li^+, which will stay in place in Ge and Si only while refrigerated.)

Other parts of a system offer more chance of failure or departure from specification. Interfaces between materials are particularly likely to be sites of change: e.g. semiconductor/air, semiconductor/metal, SiO_2/Si, metal/metal, metal/plastic. We consider further some examples of the many ways devices can fail, but we can do no more than mention human and mechanical error in manufacture, testing, and use, which together form a major fraction of the reasons for failure.

Semiconductor/air interface

Air is reactive and variable, particularly on account of the water-vapour content, so that semiconductor surfaces cannot be exposed to it if stability is to be realized. Si forms a tough, stable, adherent oxide which goes a long way to protecting the semiconductor below it. In contrast, the compounds formed by Ge with moist air are unstable, and do not form a protective film. A contaminated surface may have a high recombination velocity and a high surface conductance, so air must be sealed away from Ge surfaces. As a result, Si displaced Ge as the most widely used semiconductor and integrated circuits are made from Si not Ge. GaAs needs extra processing steps to lay down insulators on its surface.

Semiconductor/metal interfaces

The intended contact to a semiconductor will be an interface between it and a metal, though there may be a need to break down a thin layer of oxide in the process of making the contact. Failure can be either mechanical, the metal becoming detached from the semi-conductor, or electrical, when a high-resistance region forms between metal and semiconductor. Al reduces SiO_2, and has a high electrical conductivity, so is often used to make contact to Si, being deposited by evaporation in vacuum. The surface of an integrated circuit has abrupt changes of level as a result of the sequence of manufacturing operations, and one of the difficulties in applying contacts and conductor lines is the deposition of the metal as a continuous film over the steps on the surface. Another limiting process in thin conductor strips is electromigration, where the momentum of the charge carriers in a high current density can sweep atoms of the conductor along, leading to eventual disruption of the circuit.

SiO₂/Si interfaces

The temperature coefficient of expansion of SiO_2 is much less than that of Si, so on cooling after a high-temperature oxidation, the oxide layer is strained. When the oxide layer is thick, the strain may be so severe that the oxide splits from the surface and spalls off. The effect of this depends on whether it occurs during or after manufacture. A different mode of failure occurs when there are sodium ions in or on the surface of the SiO_2. Sodium is able to move readily through SiO_2, and in a situation such as the gate insulator of an n-channel MOS transistor, there are fields—either built-in or externally applied—that can drive the ions through the oxide to collect at the SiO_2/Si interface. The charge on the ions changes the gate-source bias voltage required to invert the semiconductor and create a conducting channel, so the operating point of the n-MOS circuits could drift continuously and seriously either in use or in passive storage. The more successful early MOS circuits used a p-channel, where the Na ions are driven to the metal/SiO_2 interface where they have little effect. Cures have included the exclusion of all Na from manufacturing equipment (human beings are good sources of Na!), and the use of other insulators (e.g. Si_3N_4) which are less permeable to Na ions.

Metal/metal interfaces

Metallurgical or chemical reactions between metals may result in changes in physical properties which affect device performance. Consequently, a physicist concerned with the failure of semiconductor devices must extend his interest to such topics. A brittle purple alloy formed between Au and Al at moderate temperatures surprised device technologists and led to many failures. Once the cause has been realized, ways of making reliable connection could be devised, for instance by the use of an intermediate layer of Mo or Cr.

Techniques for achieving high reliability

Some of the points mentioned below apply with special force to semiconductor manufacture, but many are relevant to the production of any reliable system.

The material from which a device is made must be of very high purity—semiconductor electronics has resulted in the development of purer materials than were required for any other purpose. In addition the containers and tools used to handle the devices must be free of the most damaging contaminants. It has been found necessary to check on the checks which suppliers make to ensure the quality of their products—merely specifying the required properties is not enough. From the reciprocal point of view, it is necessary to ensure that users have the information they need to use devices effectively and safely, and that reports of failure in use are obtained and acted on.

To infer from a failed device the cause of its failure is often a piece of skilled scientific detective work, relying on a knowledge of physics, chemistry, metallurgy, engineering, and perhaps biology and psychology. This knowledge may be embodied in one man, but is more likely to represent the combined expertise of a team.

Many of the causes of failure proceed more rapidly at high temperature. A designer using electronic components will therefore need to ensure that all the components stay cool. An example which shows how much trouble is worth taking on occasion is the use of diamond as a heat-transfer element for a high-power GaAs Gunn oscillator, because certain selected diamonds have the highest known thermal conductivity at 300 K. An alternative emphasis can be given to this point by remarking that reliability may be enhanced by derating devices so they are subject to a reduced stress.

Over a long manufacturing run the properties of a nominally standard device can vary considerably. However, the properties of two devices made from the same slice of Si, or even better the properties of two adjacent devices on the same chip are likely to be very similar. Where possible, then, devices should be balanced against each other and the performance of circuits made to depend on the relative rather than the absolute values of components.

PROBLEMS

6.1. Design an integrated circuit resistor to have a resistance of 150 Ω. The resistor is to be isolated using junction isolation, the standard doping density for both p- and n-regions is to be 10^{22} m^{-3}, and the minimum depth or width of a pattern is 5×10^{-6} m.

6.2. Check the parasitic depletion-layer capacitance of your resistor (not forgetting that it has sides), assuming that the junction has 2·5 V reverse bias, and hence find its high-frequency limit (from CR).

6.3. Take a device and list the interfaces between materials where instability could lead to malfunction of the device.

6.4. Calculate the number of memory cells that can be addressed on a chip 3 mm × 3 mm if a current of 10 mA must be allowed to flow in any X and Y address line. Al metallization of 1 μm thickness and a maximum current density of 10^9 A m^{-2} is to be used. Allow 2 μm spacing between conductors and assume two layers of metallization.

6.5. Carry out a scaling exercise for bipolar transistors in the way that was done for MOS devices in Table 6.1. Choose a way of scaling voltage on some rational basis.

Appendix: Room temperature properties of semiconductors

Material	Band gap		Lattice spacing a_0(nm)	Effective masses		Intrinsic carrier concentration n_i(m^{-3})	Electron affinity χ(eV)	Carrier mobilities		Relative permittivity ε_r	Melting point (K)
	Energy [eV]	Type		$\dfrac{m_e}{m_{e0}}$	$\dfrac{m_h}{m_{e0}}$			$\dfrac{\mu_e}{(m^2/V \cdot s)}$	$\dfrac{\mu_h}{(m^2/V \cdot s)}$		
Ge	0·66	Indirect	0·5657	0·22	0·3	2×10^{19}	4·13	0·39	0·19	16	1210
Si	1·11	Indirect	0·5431	0·97	0·5	2×10^{16}	4·01	0·15	0·06	11·8	1693
AlP	2·45	Indirect	0·5451	—	0·70	$\sim 10^{5}$	—	—	—	9	2803
AlAs	2·16	Indirect	0·5661	0·15	0·79	$\sim 10^{7}$	(2·62)	0·03	—	10·1	2013
AlSb	1·58	Indirect	0·6135	0·12	0·98	$\sim 10^{11}$	3·64	0·02	0·04	14·4	1333
GaP	2·26	Indirect	0·5451	0·82	0·60	$\sim 10^{6}$	(4·0)	0·017	0·01	11·1	1740
GaAs	1·42	Direct	0·5653	0·07	0·48	$\sim 10^{13}$	4·05	0·85	0·04	13·1	1511
GaSb	0·73	Direct	0·6096	0·04	0·44	$\sim 10^{19}$	4·03	0·40	0·14	15·7	983
InP	1·35	Direct	0·5870	0·08	0·64	$\sim 10^{14}$	(4·4)	0·4	0·015	12·4	1335
InAs	0·36	Direct	0·6058	0·02	0·40	$\sim 10^{21}$	4·54	3·2	0·05	14·6	1215
InSb	0·17	Direct	0·6479	0·014	0·40	$\sim 10^{22}$	4·59	7·8	0·4	17·7	798

Taken with permission from *Optical Communication Systems*, by J. Gower (see bibliography).

189

Answers to selected problems

1.3. $0.4, 8 \times 10^{-11}, 2.6 \times 10^{-13}$.
1.4. For electrons, 5.3×10^{-8} m or 230 interatomic spaces. For holes, 1.6×10^{-8} m or 60 interatomic spaces.
1.7. $140 \, \text{m s}^{-1}, 7 \, \mu\text{s}; 10^5 \, \text{m s}^{-1} \, (= v_{\text{max}}), 10^{-11} \, \text{s}$.
2.1. $18.6, 0.054 \, \text{A}, 1.2 \times 10^{10} \, \text{W m}^{-3}$.
2.3. 5×10^{-6} s.
2.4. $4.6 \times 10^{21} \, \text{m}^{-3}, 2.6 \, T$.
2.6. $450 \, \Omega$ per square. No, no.
3.1. $0.51 \, \text{V}, 1.12 \times 10^{-5}$ m in the n-region; 1.12×10^{-6} m in the p-region; $0.085 \, \text{pF}; 327 \, \text{V}$.
3.2. Ratio is 1.6.
3.3. 324 ns.
4.1. For the Si transistor, $\beta = 120$, $r_0 = 2.5 \, \text{k}\Omega$, $g_\pi = 8 \, \text{mA V}^{-1}$, $g_m = 1.0 \, \text{A V}^{-1}$. For the Ge transistor $\beta = 35$, $r_0 = 1.3 \, \text{k}\Omega$, $g_\pi = 15 \, \text{mA V}^{-1}$, $g_m = 1.0 \, \text{A V}^{-1}$.
      The values for g_π and g_m assume that simple theory applies validly.
4.3. $0.0027 \, \text{V}$.
4.5. For 1Ω, $4 \times 10^{-8} \, \text{A}$, $4 \times 10^{-8} \, \text{V}$; for $1 \, \text{M}\Omega$, $4 \times 10^{-11} \, \text{A}$, $4 \times 10^{-5} \, \text{V}$.
4.6. Minimum F is 1.54 at $R_s = 540$.
4.7. Band gap high to reduce leakage current. Able to be doped both n-type and p-type. Minority carrier recombination time long. Mobility high.
4.8. $3.6 \, \text{V}$.
4.9. $9 \, \text{pF}$.
4.10. $0.14 \times 10^{-12} \, C = 9 \times 10^5$ electrons.
4.11. $4 \times 10^4 \, \text{V m}^{-1}, 1.5 \times 10^{-9}$ s, holes.
5.1. $2 \times 10^{-7} \, \text{S}$ per square.
5.2. $1.4 \times 10^{-4} \, \text{C m}^{-2}$.
5.3. $4 \times 10^{-9} \, \text{A}$.
5.5. $4.5 \times 10^7 \, \text{V m}^{-1}$.
5.6. $10^3 \, \text{V m}^{-1}$.
5.7. $V_t = -0.67 \, \text{V}$. The four parts of V_t are $+0.6 \, \text{V}, +0.45 \, \text{V}, -1.0 \, \text{V}, -0.72 \, \text{V}$.
6.4. $62\,000$ bits maximum, as each line takes $12 \, \mu\text{m}$.

Bibliography

There follows a small selection (mostly recent) containing many further references.

BAR-LEV, A. (1984) *Semiconductor and electronic devices* (2nd edn). Prentice-Hall, London. Includes discussion of recent materials and devices.

CARROLL, J. E. (1974) *Physical models for semiconductors*. Arnold, London. An undergraduate text describing device physics.

GOWAR, J. (1984) *Optical communication systems* Prentice-Hall, London. Good on transmitters and detectors for optical fibres.

MEAD, C. and CONWAY, L. (1980) *Introduction to VLSI systems*. Addison-Wesley, New York. Discusses the how and why of integrated circuits. Lots of pictures.

MILNES, A. G. (1980) *Semiconductor devices and integrated electronics*. Van Nostrand Reinhold, New York. A thousand pages and thousands of references, but not for the solo tyro.

RICHMAN, P. (1973) *MOS field-effect transistors and integrated circuits*. Wiley-Interscience, New York. Good on MOS device physics.

ROSENBERG, H. M. (1978) *Solid state physics* (*OPS 9*). Clarendon Press, Oxford. Second edition. Deals more thoroughly with the material of chapter 1.

SEEC Series. (Semiconductor Electronics Education Committee) *Vols. 1–7.* (1964) John Wiley, New York. A valuable series dealing with junction transistors in detail. The section on noise in the present text follows the treatment in Vol. 4.

SMITH, R. A. (1978) *Semiconductors*. Cambridge University Press. A revised version of a valuable source-book on the underlying physics of devices.

SZE, S. M. (1981) *Physics of semiconductor devices* (2nd edn). Wiley-Interscience, New York. An advanced text with much technical information.

YANG, E. S. (1978) *Fundamentals of semiconductor devices*. McGraw-Hill, New York. An advanced undergraduate text.

Physical constants and conversion factors

Avogadro constant	L or N_a	$6 \cdot 022 \times 10^{23}\,\text{mol}^{-1}$
Bohr magneton	μ_B	$9 \cdot 274 \times 10^{-24}\,\text{J}\,\text{T}^{-1}$
Bohr radius	a_0	$5 \cdot 292 \times 10^{-11}\,\text{m}$
Boltzmann constant	k	$1 \cdot 381 \times 10^{-23}\,\text{J}\,\text{K}^{-1}$
charge of an electron	e	$-1 \cdot 602 \times 10^{-19}\,\text{C}$
Compton wavelength of electron	$\lambda_C = h/m_e c = 2 \cdot 426 \times 10^{-12}\,\text{m}$	
Faraday constant	F	$9 \cdot 649 \times 10^4\,\text{C}\,\text{mol}^{-1}$
fine structure constant	$\alpha = \mu_0 e^2 c/2h = 7 \cdot 297 \times 10^{-3}(\alpha^{-1} = 137 \cdot 0)$	
gas constant	R	$8 \cdot 314\,\text{J}\,\text{K}^{-1}\,\text{mol}^{-1}$
gravitational constant	G	$6 \cdot 73 \times 10^{-11}\,\text{N}\,\text{m}^2\,\text{kg}^{-2}$
nuclear magneton	μ_N	$5 \cdot 051 \times 10^{-27}\,\text{J}\,\text{T}^{-1}$
permeability of a vacuum	μ_0	$4\pi \times 10^{-7}\,\text{H}\,\text{m}^{-1}$ exactly
permittivity of a vacuum	ε_0	$8 \cdot 854 \times 10^{-12}\,\text{F}\,\text{m}^{-1}$ $(1/4\pi\varepsilon_0 = 8 \cdot 988 \times 10^9\,\text{m}\,\text{F}^{-1})$
Planck constant	h	$6 \cdot 626 \times 10^{-34}\,\text{J}\,\text{s}$
(Planck constant)/2π	$\hbar$	$1 \cdot 055 \times 10^{-34}\,\text{J}\,\text{s} = 6 \cdot 582 \times 10^{-16}\,\text{eV}\,\text{s}$
rest mass of electron	m_e	$9 \cdot 110 \times 10^{-31}\,\text{kg} = 0 \cdot 511\,\text{MeV}/c^2$
rest mass of proton	m_p	$1 \cdot 673 \times 10^{-27}\,\text{kg} = 938 \cdot 3\,\text{MeV}/c^2$
Rydberg constant	$R_\infty = \mu_0^2 m_e e^4 c^3/8h^3 = 1 \cdot 097 \times 10^7\,\text{m}^{-1}$	
speed of light in a vacuum	c	$2 \cdot 998 \times 10^8\,\text{m}\,\text{s}^{-1}$
Stefan–Boltzmann constant	$\sigma = 2\pi^5 k^4/15h^3 c^2 = 5 \cdot 670 \times 10^{-8}\,\text{W}\,\text{m}^{-2}\,\text{K}^{-4}$	
unified atomic mass unit (^{12}C)	u	$1 \cdot 661 \times 10^{-27}\,\text{kg} = 931 \cdot 5\,\text{MeV}/c^2$
wavelength of a 1 eV photon		$1 \cdot 243 \times 10^{-6}\,\text{m}$

$1\,\text{Å} = 10^{-10}\,\text{m}$; $1\,\text{dyne} = 10^{-5}\,\text{N}$; $1\,\text{gauss (G)} = 10^{-4}\,\text{tesla (T)}$;
$0\,°\text{C} = 273 \cdot 15\,\text{K}$; $1\,\text{curie (Ci)} = 3 \cdot 7 \times 10^{10}\,\text{s}^{-1}$;
$1\,\text{J} = 10^7\,\text{erg} = 6 \cdot 241 \times 10^{18}\,\text{eV}$; $1\,\text{eV} = 1 \cdot 602 \times 10^{-19}\,\text{J}$; $1\,\text{cal}_{\text{th}} = 4 \cdot 184\,\text{J}$;
$\ln 10 = 2 \cdot 303$; $\ln x = 2 \cdot 303 \log x$; $e = 2 \cdot 718$; $\log e = 0 \cdot 4343$; $\pi = 3 \cdot 142$

Index